GREAT IDEAS
IN PHYSICS

GREAT IDEAS IN PHYSICS

Alan Lightman

McGRAW-HILL, INC.

New York St. Louis San Francisco
Auckland Bogotá Caracas Lisbon
London Madrid Mexico Milan Montreal
New Delhi Paris San Juan Singapore
Sydney Tokyo Toronto

GREAT IDEAS IN PHYSICS

1 2 3 4 5 6 7 8 9 0 DOC DOC 9 0 9 8 7 6 5 4 3 2 1

ISBN 0-07-037935-1 (hard cover)
ISBN 0-07-037937-8 (soft cover)

This book was set in Palatino by The Clarinda Company.
The editors were Susan J. Tubb and Jack Maisel;
the production supervisor was Annette Mayeski.
The cover was designed by Joan Greenfield.
R. R. Donnelley & Sons Company was printer and binder.

Library of Congress Cataloging-in-Publication Data

Lightman, Alan P., (date).
    Great ideas in physics / Alan Lightman.
        p.   cm.
    ISBN 0-07-037935-1 (hard cover)—ISBN 0-07-037937-8 (soft cover)
    1. Physics.   I. Title.
    QC21.2.L54   1992
    530—dc20   91-27027

Contents

Excerpts

GREAT IDEAS
IN PHYSICS

Introduction

Several years ago, I went to Font-de-Gaume, a prehistoric cave in France. The walls inside are adorned with Cro-Magnon paintings done 15,000 years ago, graceful drawings of horses and bison and reindeer. One particular painting I remember vividly. Two reindeer face each other, antlers touching. The two figures are perfect, and a single, loose flowing line joins them both, blending them into one. The light was dim, and the colors had faded, but I was spellbound.

Likewise, I am spellbound by the plays of Shakespeare. And I am spellbound by the second law of thermodynamics. The great ideas in science, like the Cro-Magnon paintings and the plays of Shakespeare, are part of our cultural heritage.

A painter paints a sunset, and a scientist measures the scattering of light. The beauty of nature lies in its logic as well as appearance. And we delight in that logic: The square of the orbital period of each planet equals the cube of its distance from the sun; the shape of a raindrop is spherical, to minimize the area of its surface. Why it is that nature should be logical is the greatest mystery of science. But it is a wonderful mystery.

Discoveries in science are not just about nature. They are about people as well. After Copernicus, we have taken a more humble view of our place in the cosmos. After Darwin, we have recognized new relatives clinging to the family tree. The great ideas of science have changed our view of the world and of ourselves. Science is a human activity as well as an exploration of nature, and, as a human activity, science connects to philosophy, history, literature, and art.

These, then, are the two aims of this book: to provide a grasp of the nature of science, and to explore the connections between science and the humanities. Our exemplary science will be physics.

No attempt has been made to provide a survey of all physics. Instead, this book has been organized around a small number of *ideas*. The ideas are the conservation of energy, the second law of thermodynamics, the relativity of time (relativity theory), and the wave-particle duality of nature (quantum theory). Each of these landmark ideas has changed our world view. Each has had impact and application far beyond science. The law of the conservation of energy, which deals with an indestructible property of nature, provides a fundamental example of the logic and predictive power of science. The second law of thermodynamics, which states that all isolated physical systems unavoidably become more disordered in time, explains why machines cannot keep running forever. The relativity of time, Einstein's discovery, states that time does not flow at an absolute rate, as it seems, but depends on the motion of the clock or observer. Relativity theory shows that our instincts about nature may sometimes be wrong. The quantum theory, which states that objects behave as if they were in two places at once, requires a new conception of reality.

In discussing the physical world, we will learn something of the scientific method. Most science uses inductive reasoning: the scientist makes a number of observations of nature, finds a pattern, generalizes the pattern into a "law" or organizing principle, and then tests that law against future experiments. The discovery of the law of the conservation of energy is an example of inductive science. Deductive science is more rare. Here, the scientist begins by postulating certain truths of nature, with little guidance from outside experiment, and deduces the consequences of those postulates. The consequences are cast into predictions, which can then be pitted against experiment. The theory of relativity is an example of deductive science. Both inductive and deductive reasoning in science are "scientific" in that theories are ultimately judged by their agreement or disagreement with experiment.

In discussing the physical world, we will also encounter a number of approximations and simple models of nature: collisions between balls on frictionless tables, pendulums swinging in three-molecule gases, and so on. Approximations and models are crucial to science. A scientific model begins with a real physical object or system, replaces the original object with a simpler object, and then represents the simplified object with equations describing its behavior. Like a toy boat, a scientific model is a scaled-down version of a physical system, missing some parts of the original. Deciding what parts should be left out requires judgment and skill. The omission of essential features makes the model worthless. On the other hand, if nothing is left out, no simplification has been made and the situation is often too difficult to analyze. In making a model of a swinging pendulum, for example, we might at first try to include the detailed shape of the weight at the end, the density and pressure of the air in the room, and so on. Finding such a description much too complex to manage, we could approximate the weight by a round ball and neglect the air completely. This much simpler system behaves much more like the original. But if we left out gravity, the resulting theoretical pendulum would not swing back and forth. By solving the equations of a model, predictions can be made about the original physical system and then tested.

Alas, the equations. The language of science is mathematics, and it is impossible to appreciate science without equations and quantitative problems.

(I will discuss the role of mathematics in science in the first chapter, in the section on gravitational energy.) Quantitative problems and solutions are scattered throughout the text. However, only high school mathematics, without calculus, is required. An Appendix reviews all of the math you will need. The equations and problems may seem demanding and tedious, but they are well worth the effort. You will not find a watered-down treatment of science in this book. If you invest the time, you will get the real thing.

The other dimension of the book is the humanistic. The relationship between science and the humanities is two-way. Science changes our view of the world and our place in it. In the other direction, the humanities provide the store of ideas and images and language available to us in understanding the world. The exploding star of A.D. 1054, the Crab Nebula, was sighted and documented by the Chinese, but nowhere mentioned in the West, where the Aristotelian notion of the immortality of stars still held sway. We often do not see what we do not expect to see.

The humanistic sources and effects of the ideas in this book are suggested by readings and excerpts from the original scientific literature and from history, literature, philosophy, and art. For example, readings from the early Roman poet Lucretius show the psychological comfort brought by a conservation law; readings from historian Henry Adams show his attempt to apply the second law of thermodynamics to an understanding of the decline of human civilization; readings from Einstein's autobiography show the influence of the philosopher David Hume on Einstein's formulation of the theory of relativity; excerpts from a novel by Vladimir Nabokov illustrate the use of ideas from relativity in literature; readings from the philosopher Ernst Cassirer consider the implications of quantum physics for human ethics.

The humanities excerpts in the text are brief, by necessity, and should be considered only as starting points for further reading. Furthermore, I have taken a rather conservative approach in discussing the humanities. I have included only those excerpts and references that show a direct connection to science. In fact, many of the important connections are indirect and diffuse. New ideas in science, literature, philosophy, and art are often part of a broad change in thinking about the world, and it is impossible to identify that change with a single phrase, such as "the relativity of time," or to decide where an idea started. Nevertheless, I have stuck to examples where the connections between ideas in science and in the humanities are clear and explicit. It seems good to start with examples we can be sure about and then build on those.

Discussion questions will challenge you to think broadly about the ideas and to relate the science to your own interests. The discussion questions are not answered to avoid oversimplification and to allow you to ponder the issues without restraint.

The material in each chapter includes a treatment of the scientific content of the idea, with demonstrations and activities where possible; quantitative problems with solutions; a reading list from the humanities and from the original scientific literature; excerpts from these readings; and discussion questions probing the human dimensions and applications of the ideas. Experiments with nature and a first-hand experience with the scientific method are critical elements of any study of science, and a separate laboratory is included in the course.

For The Instructor

Recent national studies have uncovered a startling illiteracy in science, and new approaches to the teaching of science seem to be called for. This is such an attempt—an interdisciplinary course in science for the nonscience major, centered on a small number of ideas.

Several concerns have determined the style of the book. The material should be accessible to the student with little science background, while preserving the depth and integrity of the subject. To this end, care has been taken to reduce each idea to its essentials, with minimum terminology, and to use only high school mathematics, without calculus. Given these limitations, the essentials are treated with rigor. The emphasis is placed on concepts rather than facts, and the science connects to the body of knowledge and orientation of the nonscience student, a connection not usually made in traditional courses. Finally, each concept should be developed in enough detail so that a solid understanding emerges, with perhaps a new, self-consistent model of the world. This requirement, by necessity, limits the course to a small number of ideas, as opposed to the array of many topics found in a survey course.

The course develops only those concepts needed to understand the four central ideas treated. Many standard concepts in physics, such as the concept of force and the entire field of electricity and magnetism, have been largely omitted.

The book is designed for a one-semester course, with roughly 4 weeks on each chapter. Such a course might alternate science lectures with discussion sessions on the readings, giving equal time to each.

The discussion questions throughout the text are intended for open classroom discussion as opposed to formal lectures. Some of the discussion questions can also be used for essay assignments. In most cases, I have provided little of my own commentary on the excerpts and leave the interpretations and connections to be drawn out of the discussion questions. In the few cases where I do give a brief commentary, my comments should not be taken as gospel. It would be best for students to read the complete works from which the excerpts are taken, as time permits.

The problems and solutions scattered throughout the text are intended not as homework, but as amplifications of the text and as examples of quantitative applications of the ideas. These problems vary greatly in difficulty, and it is expected that students will need to work slowly through the more difficult solutions. For homework, instructors may want to design their own problems, perhaps similar to some of the examples in the book.

The most technically difficult part of the book is the derivation of the Heisenberg uncertainty principle, in Section IV–F. In light of the fundamental importance of the uncertainty principle, I decided to include this section in full detail so that readers could see for themselves where the principle comes from. However, some readers may want to traverse this section quickly, without trying to follow it in detail, and others may want to skip it entirely, going directly to the result in Eq. (IV–28). In the latter case, momentum must first be defined.

Finally, the demonstrations and experiments discussed in Chapters I and II require only simple equipment and can easily be done in class. The laboratory, which requires about two hours, can be done at any time of the course.

Acknowledgments

This book is based on a new course I taught in 1988 to undergraduates, mostly freshmen, in the Council of Humanities at Princeton University. I am grateful to Carol Rigolot, then associate director of the Council of the Humanities, and to David Wilkinson, then chairman of the Princeton physics department, for their warm encouragement and support. I am also grateful to Irwin Shapiro and the Harvard-Smithsonian Center for Astrophysics for their encouragement and support of the project during its initial stages. Owen Gingerich and Gerald Holton also provided welcome encouragement at the beginning. At M.I.T., where I have completed most of the work on the book, I thank Kenneth Manning and the School of Humanities, Arts, and Social Sciences, and the physics department, for their encouragement of my interdisciplinary interests. For helpful comments on the manuscript, I thank Morton Brussel, Robert Clark, Robert Di Yanni, Anthony French, Katherine Hayles, Hans Hoch, Renate Holub, Allison Oppenheimer, John Rigden, Harriet Ritvo, Harold Rorschach, David Rowe, George Rybicki, Lawrence Shepley, Claire Silvers, Jearl Walker, and Steve Weininger. I am particularly grateful to Irene Nunes for her intelligent and thorough reading of the manuscript. I thank my editors at McGraw-Hill, Denise Schanck and Susan Tubb, for their receptiveness to new ideas and their encouragement.

JULIUS ROBERT MAYER

Robert Mayer (1814–1878), the son of the owner of an apothecary shop, was born in Heilbronn, Württemberg (now Baden-Württemberg), Germany. His two older brothers went into their father's business, while Robert went to medical school at the University of Tübigen. In 1837, he was expelled from the university for his membership in a secret student society, but he was readmitted the following year and received his doctorate of medicine in 1838.

Mayer spent the year of 1840 as a physician aboard a Dutch merchant ship in the East Indies. It was here, by a curious route, that he was led to the idea of the conservation of energy. That route involved medicine, not physics. Letting out blood was a common medical cure of the time, and while letting the blood of sailors arriving at Java, Mayer noted that their blood was unusually red. He reasoned that the heat of the tropics reduced the metabolic rate needed to keep the body warm and therefore reduced the amount of oxygen that needed to be extracted from the blood. Accordingly, the sailors had a surplus of oxygen in their blood, causing its extra redness. This hypothesis, and its apparent validation, were taken by Mayer to support the link between heat and chemical energy, the energy released by the combustion of oxygen. After deciding that there must be a balance between the input of chemical energy and the output of heat in the body, Mayer made a conceptual leap. Friction in the body, from muscular exertion, also produced heat, and the energy associated with this heat also had to be strictly accounted for by the intake of food and its content of chemical energy.

Mayer, being a physician and not a physicist, was at first not familiar with the principles of mechanics, and his first paper on energy had errors. It was rejected. Although disappointed, Mayer immediately took up the study of physics and mathematics, learned about kinetic energy, and submitted a new paper a year later, in 1842. The new paper stated the mechanical equivalent of heat. Like the previous one, the new paper also argued that energy is a cause, that every cause has an effect, and that energy is indestructible. Such notions represented an enormous extrapolation from Mayer's actual observations. Indeed, Mayer conducted very few experiments in his scientific career.

Although Mayer's second paper was published, and later celebrated, it was given little attention at the time. Mayer's success was partly limited by his lack of a traditional background in physical science, his corresponding "outsider" status, and his lack of institutional affiliations. He was principally a medical doctor, practicing in his native town of Heilbronn. Despondent over his lack of recognition, Mayer attempted suicide in 1850. He suffered episodes of insanity in the early 1850s and was confined in asylums on several occasions.

After 1860, Mayer was finally given the recognition he deserved. Many of his articles were translated into English, and such well-known scientists as Rudolph Clausius in Germany and John Tyndall in England began to champion Mayer as the founder of the law of the conservation of energy. Unfortunately, Mayer had little impact during the time of his work. Almost all of his ideas were later rediscovered by other scientists much better known and recognized, such as James Joule, Karl Holtzmann, and Hermann von Helmholtz.

From his marriage to Wilhelmine Regine Caroline Closs in 1842, Mayer had 7 children, 5 of whom died in childbirth. Mayer died of tuberculosis in 1878.

CHAPTER 1

The Conservation of Energy

Here hills and vales, the woodland and the plain,
Here earth and water seem to strive again,
Not chaos-like together crushed and bruised,
But, as the world, harmoniously confused:
Where order in variety we see,
And where, though all things differ, all agree.
Alexander Pope, *Windsor Forest (1713)*

A. CONSERVATION LAWS

In volume I of *The Feynman Lectures on Physics*, Richard Feynman gives a delightful example of a conservation law. A mother gives her child 28 blocks to play with. They are sturdy blocks and cannot be split or chipped. One day, the mother comes home and sees only 26 blocks. She anxiously looks around the house and eventually finds the two missing blocks under the bed. The next day, only 24 blocks can be found. She looks under the bed and everywhere else, but she cannot turn up any more blocks. Then she shakes her child's little locked box and hears a rattle. She knows that the box, when empty, weighs 16 ounces, and she also knows that each block weighs 2 ounces. She puts the box on a scale, and it registers 24 ounces. Aha! That accounts for the four missing blocks. One morning a week later, the mother sees only 8 blocks. She finds 3 blocks under the bed and 2 in the box, making a total of 13, but no more anywhere in the house. Then she notices that the water in the fish bowl is dirty, and it seems a little high. The water level is normally 12 inches. She measures it now and finds it to be 13.5 inches. She carefully drops a block in the bowl and discovers that the water level goes up 0.1 inch. Aha! If one block pushes the water level up 0.1 inch, and the level is up 1.5 inches higher than usual, there must be 15 blocks in the bowl. That accounts for all the missing blocks.

After several weeks, the mother decides that she has found all the hiding places. To keep track of the blocks seen and unseen, she works out the following rule:

$$(\# \text{ blocks seen}) + (\# \text{ blocks under bed}) + \frac{(\text{weight of box}) - 16 \text{ ounces}}{2 \text{ ounces}}$$
$$+ \frac{(\text{height of water}) - 12 \text{ inches}}{0.1 \text{ inches}} = 28 = \text{constant.} \qquad (I\text{–}1)$$

This equation is an abbreviated but precise expression of what the mother has learned. A little later I will have more to say about the role of equations and mathematics in science.

The mother has discovered a *conservation law:* The total number of blocks remains constant, although it may take a little ingenuity to find them all. From one day to the next, the blocks may be parceled out and hidden in a variety of ways, but a logical system can always account for all 28.

All conservation laws are like this one. Some countable quantity of a system never changes, even though the system may undergo many changes. A conservation law does not reveal how a system operates or what causes it to change. Still, such a law can be extremely helpful for making predictions. For example, the mother could use her rule to figure out how many blocks would be in the box if she knew how many were under the bed and how many were in the fish bowl. From the known, she can reach to the unknown.

In most conservation laws of nature, the "blocks" are never visible. Counting them is therefore an indirect process, analogous to weighing the box and measuring the height of the water. Moreover, the conserved quantity is usually not a tangible object, like a block, but something more abstract. Nevertheless, all conservation laws express the result that some quantifiable property of a system never changes.

In this chapter we will study the conservation of energy. *Energy is the capacity to do work.* A moving hammer, a coiled spring, a paperweight on a desk, and boiling water all represent different forms of energy. A moving hammer can push a nail into a wall; a coiled spring can turn the hands of a clock; a falling weight can lift another weight or move the paddles of an electric generator and light up a house; boiling water can cook an egg. Despite the many forms that energy can take, scientists have found that the total energy of an *isolated system* never changes. An isolated system is an object or group of objects that has no contact with the surrounding world. The conservation, or constancy, of energy is a strange and beautiful fact about nature. Nature possesses other conservation laws, such as the conservation of momentum and the conservation of electrical charge, but for brevity we will restrict our attention to the conservation of energy, which well illustrates the meaning and power of all conservation laws. Why it is that any conservation laws exist is a mystery. If there were no conservation laws, the predictive power of science would be much reduced. The world would seem a far more irrational place.

Later in the chapter, using historical examples, I will suggest that conservation laws reflect a deep-seated human desire for order and rationality. That is not to say that human beings invented conservation laws. Nature invented conservation laws and lives by them. But the *idea* of a conservation law, of some permanent and indestructible quantity in the world, was conceived by humankind long ago and appeals to our minds and emotions.

B. GRAVITATIONAL ENERGY

Energy comes in many forms. There is gravitational energy, kinetic energy (energy of motion), heat energy, elastic energy, electrical energy, nuclear energy, and so on. We will have to discover the rule for how to count each kind of energy, just as Eq. (I–1) provides a rule for counting the child's blocks in each hiding place. When we have figured out all the rules, the law of the conservation of energy will look like

$$E_G + E_K + E_H + E_E + \cdots = \text{constant}, \qquad (I-2)$$

where E_G stands for gravitational energy, E_K stands for kinetic energy, and so on. In this chapter, we will limit ourselves to gravitational energy, kinetic energy, and heat energy, which will be sufficient to acquaint us with the meaning and application of the conservation of energy. Physics is quantitative. We now have to roll up our sleeves and discover the rules for measuring the different kinds of energy. We will begin with gravitational energy.

1. A Demonstration with an Inclined Plane

In determining how to quantify gravitational energy, it is best to start with a system in which *only* gravitational energy is involved. Such a system is shown in Fig. I–1. Place an inclined plane (basically a small ramp) on a table, hanging over the edge. Mount a pulley at the top of the inclined plane, from which are hung two different masses, indicated by m_1 and m_2. The plane should be as smooth as possible, so that the mass m_2 can easily slide along its surface. In Figure I–1a, we have chosen an inclined plane with a slope length to height in the ratio of 5 to 3.

The masses m_1 and m_2 should be made of different numbers of the same unit mass. The unit mass can be anything—a penny, for example. Mass is simply the quantity of matter. Six pennies have twice the mass of three pennies, and so on. In the figure, m_2 is 5 units of mass and m_1 is 3. Whatever the unit of mass, m_2 has 5/3 as much mass as m_1.

Set up an experiment as in Fig. I–1a, with the same 5 to 3 proportions for the inclined plane and the same ratio of masses, $m_2/m_1 = 5/3$. Now give the

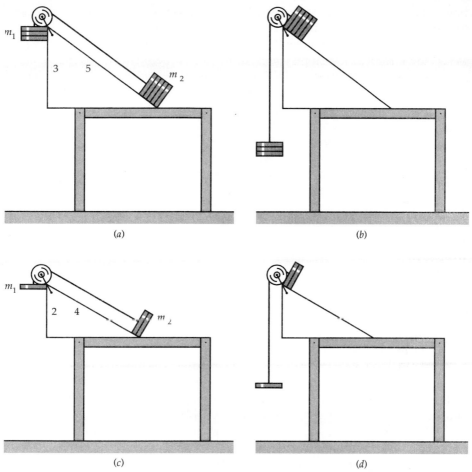

Figure I–1 (*a*) A right-angled inclined plane, with the masses m_1 and m_2 connected by a string passing through a pulley at the top of the plane. (*b*) The mass m_1 has fallen as low as possible, raising the mass m_2. (*c*) and (*d*) Repetitions of (*a*) and (*b*) for other inclined planes and masses.

mass m_2 a slight upward nudge. It will slowly slide up to the pulley while m_1 slowly falls down, as shown in Fig. I–1b. The slight initial nudge is needed because there will always be at least a tiny bit of friction between the mass and the plane, and this friction must be overcome. Try increasing the ratio m_2/m_1 by adding to the mass m_2. Now the mass m_2 will not be able to slide up the plane without constant pushing. If, instead, we decrease the ratio m_2/m_1 by adding to the mass m_1, m_1 will spontaneously fall down on its own, moving faster and faster until it reaches the bottom of its fall. (A note of caution: In any realistic experiment, there is a small amount of friction, which will cause slight discrepancies between the actual experimental results and the results quoted here. The discrepancy diminishes with less friction and is almost unnoticeable with the equipment available in college science labs.)

Summarizing our observed results so far, when the masses m_2 and m_1 are in the same ratio as the slope length to height of the plane, $m_2/m_1 = 5/3$, and only in that ratio, the system in Fig. I–1a can slowly change to the system in Fig. I–1b without any constant pushing from the outside and without any spontaneous and increasing speeds of the masses. Only a tiny nudge is needed to make the system change.

Do a second experiment, varying the proportions of the inclined plane. This time, use an inclined plane whose slope length to height is in the ratio of 4 to 2, as shown in Figs. I–1c and I–1d. Now you will find that for $m_2/m_1 = 5/3$ the masses are unbalanced and immediately move on their own, but for $m_2/m_1 = 2$, the system can slowly change from the situation of Fig. I–1c to that of Fig. I–1d. Evidently, for each set of proportions of the inclined plane, there is one particular ratio of m_2/m_1 that allows the two masses to move slowly to a new configuration without constant outside pushing and without sponta- neous motions.

To make further progress, let's think carefully about what happens in the two experiments. In the first, a mass of 5 rises a vertical height of 3 (the verti- cal distance from the top to the bottom of the inclined plane), while a mass of 3 falls a height of 5 (the length of the string connecting the masses, or the length of the slope). In the second experiment, a mass of 2 rises a height of 2, and a mass of 1 falls a height of 4. A quantitative pattern is emerging. The amount of rising mass multiplied by the vertical distance it rises equals the amount of falling mass multiplied by the vertical distance it falls: $5 \times 3 = 3 \times 5$ in the first experiment and $2 \times 2 = 1 \times 4$ in the second.

It will be extremely useful to express this result in a general form, using some symbols. (The rationale for doing so will be discussed shortly, at the end of the section on gravitational energy. A brief review of symbolic notation and the basic principles of algebra can be found in Appendix A.) Let the initial ver- tical height of mass m_1 be denoted by h_{1i} and its final vertical height be h_{1f}; let the initial vertical height of mass m_2 be h_{2i} and its final vertical height be h_{2f}. Then, the mass m_1 falls a distance $h_{1i} - h_{1f}$ and the mass m_2 rises a distance $h_{2f} - h_{2i}$. (Remember that h_{1i} is bigger than h_{1f}, and h_{2f} is bigger than h_{2i}.) Using this notation, the result we have found is

$$m_1 \times (h_{1i} - h_{1f}) = m_2 \times (h_{2f} - h_{2i}). \tag{I–3a}$$

Of course, we have proved this result for only a small number of situations. It is important to repeat the experiments for many other masses and inclined planes in order to test the generality of Eq. (I–3a). When we do, we find that it always correctly predicts the results, provided the friction is small. Such experiments and results, in somewhat different form, were known by the ancient Greeks.

Problem I–1: Lifting Elephants with Babies

A seesaw allows one mass to go up and one to come down, much like an inclined plane with a pulley. Suppose a 10-pound baby is at one end of a seesaw and a 10-ton elephant at the other, as shown in Fig. I–2a. If the baby is at the right height above the ground, she can just counterbalance the elephant, meaning that with a tiny nudge, she can slowly descend to the ground while the elephant slowly rises, with no outside pushing and no spontaneous and increasing motions. If this situation is achieved and the elephant rises 1 inch, as shown in Fig. I–2b, how far does the baby fall? In other words, what is the initial vertical height of the baby?

Solution: We can apply Eq. (I–3a), with the baby being the mass m_1 and the elephant being m_2. If we let the baby have 10 units of mass, $m_1 = 10$, then $m_2 = 20{,}000$. Measure all heights from the ground. Then the initial height of the elephant is $h_{2i} = 0$ inches, and its final height is $h_{2f} = 1$ inch. The final height of the baby is $h_{1f} = 0$ inches. We want to find the initial height of the baby, h_{1i}. Substituting these numbers into Eq. (I–3a), we get $10 \times h_{1i} = 20{,}000 \times 1$ inch or $h_{1i} = 2000$ inches, which is about 167 feet. So, a 10-pound baby dropping 167 feet can lift a 10-ton elephant 1 inch.

Figure I–2 (*a*) Baby and elephant balancing each other on a seesaw. The baby is initially at a height h_{1i}. (*b*) The baby has dropped to the ground and the elephant is now at a height of 1 inch. (Drawings not to scale.)

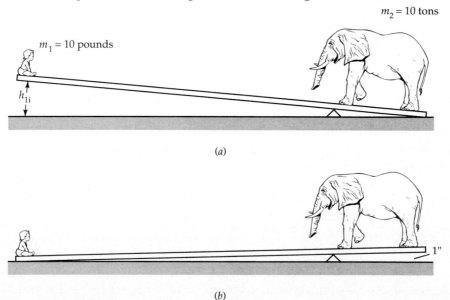

$m_2 = 10$ tons

$m_1 = 10$ pounds

h_{1i}

(*a*)

1"

(*b*)

14

The seesaw is a lever, which was quantitatively understood in ancient times by the Greek mathematician and scientist Archimedes (287–212 B.C.). The workings and balance of levers were often rediscovered through history. For example, in his notebooks, Leonardo Da Vinci (1452–1519) frequently states a result equivalent to Eq. (I–3a), in its application to the balance of a seesaw. In his *Codex Arundel,* he says, "Multiply the longer arm of the balance by the weight it supports and divide the product by the shorter arm, and the result will be the weight which, when placed on the shorter arm, resists the descent of the longer arm, the arms of the balance being above all basically balanced." The "balance" here is the seesaw, or lever. (Although Leonardo uses the word "weight" here, we can substitute "mass." Later we will discuss the relation between weight and mass.) To put Leonardo's statement in other terms, notice that the ratio of the long arm of a seesaw to the short arm is the same as the ratio of the height the light mass drops divided by the height the heavy mass rises, since the seesaw forms two similar right triangles. (See Appendix A.) Thus, Leonardo's statement in our language is

$$\frac{m_1 \times (h_{1i} - h_{1f})}{h_{2f} - h_{2i}} = m_2,$$

which is simply a rearrangement of Eq. (I–3a), obtained by dividing both sides by $(h_{2f} - h_{2i})$.

Leonardo was exquisitely sensitive to geometrical relationships, in science as well as in art. He thought of his experimental result not in terms of a conservation law, but rather as a balance of forces. The idea of the conservation of energy, and even a full understanding of energy, were not grasped until the nineteenth century.

2. The Rule for Measuring Gravitational Energy

Equation (I–3a) expresses an interesting, quantitative relationship between masses and their heights for a system undergoing change. We can rearrange the equation by putting all the "initial" terms on one side and all the "final" terms on the other. Because this is the first example of algebraic manipulations of equations, we will give all of the intermediate steps. Hereafter, some of the intermediate steps will be left out. You should consult Appendix A if you need to review the rules of algebra.

First, carry out the multiplications of m_1 on each of the terms inside the parentheses, h_{1i} and h_{1f}, and do the same for m_2:

$$m_1 \times h_{1i} - m_1 \times h_{1f} = m_2 \times h_{2f} - m_2 \times h_{2i}.$$

Let's put all the initial quantities on the left-hand side of the equation and all the final quantities on the right. We want to get the $m_1 \times h_{1f}$ to the right-hand side of the equation, so let's add it to both sides:

$$m_1 \times h_{1i} - m_1 \times h_{1f} + m_1 \times h_{1f} = m_2 \times h_{2f} - m_2 \times h_{2i} + m_1 \times h_{1f},$$

or, since the second and third terms on the left-hand side add up to zero and cancel each other,

$$m_1 \times h_{1i} = m_2 \times h_{2f} - m_2 \times h_{2i} + m_1 \times h_{1f}.$$

We want to get the $m_2 \times h_{2i}$ to the left-hand side of the equation, so let's add it to both sides:

$$m_1 \times h_{1i} + m_2 \times h_{2i} = m_2 \times h_{2f} - m_2 \times h_{2i} + m_1 \times h_{1f} + m_2 \times h_{2i},$$

or, since the second and fourth terms on the right-hand side add up to zero and cancel each other,

$$m_1 \times h_{1i} + m_2 \times h_{2i} = m_2 \times h_{2f} + m_1 \times h_{1f}.$$

Finally, we can switch the two terms on the right-hand side, obtaining

$$m_1 \times h_{1i} + m_2 \times h_{2i} = m_1 \times h_{1f} + m_2 \times h_{2f} \tag{I–3b}$$

Equation (I–3b) has the form of a conservation law. A certain quantity of the system, namely the *sum* of each mass multiplied by its height, has the same final value as initial value. Despite the changes in the system as the masses go up and down, this sum remains constant. We have discovered a conservation law.

Although Eq. (I–3b) may seem abstract and theoretical, it did not leap out of our heads. It is an *observational* result. Admittedly, we generalized from a small number of experiments, but this is often how science works. A number of experiments are done, a pattern is found, and the pattern is recast into general form. Then that general form is tested against new experiments.

Laws of nature are different from the laws of a society. Although physicists argue with themselves over whether humankind will ever fully know nature's "truth," they agree that a fundamental logic governs the physical world. Although that logic may please us, it seems to exist independently of the human mind and desires. Accordingly, the laws of nature cannot be arrived at by consensus or decree, as can statutory laws. They must be discovered by experiment and analysis. Since it is impossible to know whether we have complete knowledge of nature, we must regard the conservation of energy and all other "laws of nature" as simply our best current understanding of the world, subject to revision as warranted by new physical evidence.

Returning to the law of the conservation of energy, exactly what is being conserved in Eq. (I–3b)? Although we can happily use this equation without further ado, we will give the conserved quantity a name: gravitational energy. Gravity is involved because it is the gravitational pull of the earth that causes a mass to fall if dropped, and Eq. (I–3) deals with falling and rising masses. Equation (I-3b) says that gravitational energy is conserved for the systems we have considered.

Notice that the gravitational energy of a system is made up of several parts, one for each mass. We can call each of these parts the gravitational energy for that mass. For example, the gravitational energy of mass m_1 is $m_1 \times h_1$, and so on. This is the quantitative measure of the gravitational energy of mass m_1. (The initial value of this energy would be $m_1 \times h_{1i}$, and the final value would be $m_1 \times h_{1f}$.) When a mass changes its height, its gravitational energy changes. We can restate the conservation law by saying that the *total* gravitational energy of an isolated system of masses is conserved, where the total is understood to be the sum of the gravitational energies of the separate masses. During a change in the system, a mass that rises gains gravitational energy, and a mass that falls loses gravitational energy, but the *net change in gravitational energy is zero*. That is the meaning of a conservation law. Moreover, there could be many more than two masses in the system. Even in the systems we have considered, we could imagine that each mass is divided into a few parts, at the same locations but relabeled as separate masses. Then Eq. (I–3b) would still apply, as long as we added in the gravitational energy, $m \times h$, for every mass.

Our rule for quantifying gravitational energy is almost complete. We have tentatively identified the product of mass and height, $m \times h$, as the gravitational energy of a mass m. But we could have defined gravitational energy to be half this product or any number multiplied by $m \times h$ and our results would be the same. For example, if gravitational energy were defined to be $2 \times m \times h$, Eq. (I–3b) would still say that gravitational energy is conserved, because that equation is still true if we multiply both sides by 2. The precise number that multiplies $m \times h$ to define gravitational energy is partly a matter of convention. To allow for this flexibility, we will define the gravitational energy E_G of a mass m at height h to be

$$E_G = C \times m \times h, \tag{I–4}$$

where C stands for a constant, that is, a number that is the same for every part of the system. In a later section we will specify the constant C. It will depend on the units we choose to measure energy, mass, and height. It will also depend on the strength of gravity, and this seems logical. A mass m at a certain height h above the earth should have more gravitational energy than the same mass at the same height above the moon, since the earth's gravity is stronger than the moon's.

At this point the reader might ask how the height h is to be measured. Should it be measured from the ground, from a table, or from where? Because only *changes* in height really matter, an object's height can be measured from any place of reference as long as it is done so from the same place before and after a change.

A final clarification. Having established that the total gravitational energy is conserved for masses on inclined planes, we must remember that we intentionally set up our experiments so that *only* gravitational energy would be involved. In particular, our observational result, Eq.(I–3b), did not apply when we had to give one of the masses a continued push to get it up the inclined plane, nor did it apply when the masses spontaneously changed positions with increasing speed. In the first of these situations, our system was not isolated; it received energy from the outside world in the form of a continued push. In the second, kinetic energy was present in the form of the motions of the masses when m_1 was too heavy. In such situations, the total gravitational energy is not conserved. Gravitational energy is only one form of energy, analogous to one hiding place for the child's blocks. In general, the changes in a system will involve several different forms of energy, and we will find that it is the total of all forms of energy, not just the total gravitational energy, that is conserved. Nevertheless, it was useful to begin with a simple system involving only gravitational energy so that we could discover the rule for measuring gravitational energy and could realize that something was conserved.

3. A Mental Experiment

We did physical experiments to discover the rule for measuring gravitational energy, Eq. (I–4). However, it is possible to uncover that rule without actual experiments, with only imaginary experiments, if we *begin* with the assumption that the gravitational energy of a system is conserved and that the gravitational energy of each mass of the system can change only when its vertical height changes. Such a mental experiment, due to Richard Feynman, is shown in Fig. I–3. Imagine putting three blocks on one end of a seesaw and one block on the other end, as shown in Fig. I–3a. Place the seesaw pivot so that the blocks just counterbalance each other. The four blocks are all identical, so that the mass m_1 at the left is exactly one-third of the mass m_2 on the right, $m_1 = m_2/3$. Since equipment is cheap, let's make the blocks cubes, 1 foot on a side. If the seesaw is just balanced, it can slowly move to the position shown in Fig. I–3b, with no constant outside pushing and no spontaneous motions. We can arrange things so that the blocks on the right are lifted exactly 1 foot (one block's width) as the block on the left descends to the ground. If the blocks are lifted more than this, simply imagine moving the pivot closer to them, also moving m_1 closer to maintain balance, until the three blocks are lifted the amount we want. The question is: How far does the mass m_1 fall? We have labeled this unknown x in Fig. I–3a.

To answer the question, first dispense with the seesaw, which is only a tool for moving the masses around. Slide mass m_1 horizontally to the right until it just fits under the three blocks m_2, as shown in Fig. I–3c. This sliding doesn't change the height of any blocks, so, by assumption, it doesn't change the energy of the system. Now slide the top block horizontally to the left, as shown in Fig. I–3d, again without any change in energy of the system. The top

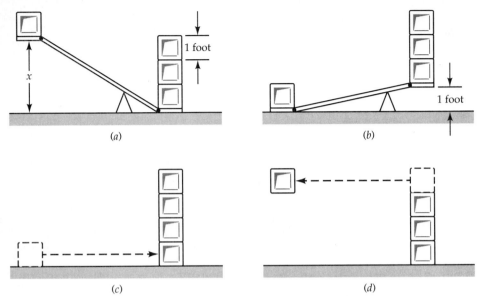

Figure I–3 (*a*) Blocks balancing each other on a seesaw. The block on the left is a height *x*. (*b*) The block on the left has dropped to the ground and the three blocks on the right have risen by 1 foot. (*c*) The block on the left is moved horizontally underneath the three blocks on the right. (*d*) The top block is moved horizontally to the left.

block is 3 feet off the ground, since each of the three blocks previously underneath it is 1 foot high. Since we have assumed that gravitational energy is conserved and since we have not added any energy to the system in any of its changes, the total gravitational energy of the systems in Figs. I–3a and I–3d must be the same. The heights of the three blocks on the right are clearly the same in Figs. I–3d and I–3a. Therefore, the block on the left must be the same height in the two figures. We know that the block on the left in Fig. I–3d is 3 feet high. Therefore, the block on the left in Fig. I–3a is also 3 feet high, or *x* = 3 feet.

In conclusion, when a mass of three units rises one unit of height, a mass of one unit falls three units of height. This is exactly what we would have predicted from our experimentally determined result, Eq. (I–3a). It is easy to see that if we repeat our imaginary experiment with different numbers of blocks on each side, we will get the general result given in Eq. (I–3). Remember that we did have to *assume* the conservation of gravitational energy from the beginning. Otherwise, there would be no reason to think that the block on the left had the same height in Figs. I–3a and I–3d.

Astoundingly, important new results in science have sometimes been found in this way, using only mental experiments. Such experiments are especially vital when the required conditions of an actual experiment cannot be duplicated on earth. In any case, mental experiments are cherished by physicists, particularly theoretical physicists, who can cause damage in the lab.

Discussion Questions I–1

Which do you find more satisfying, physical or mental experiments? What are the advantages and disadvantages of each? Which do you trust more, and why?

---

4. The Role of Mathematics in Science

At this point, it seems worthwhile to take a short detour and discuss the role of mathematics in science. In the example of the mother and her child's blocks, why did we bother to express her conclusions in the form of an equation, Eq. (I–1)? Likewise, in the last example of the rising and falling masses, why did we bother to express what we had learned in the form of an equation Eq. (I–3a), using the symbols m_1 and h_{1i}? Why don't we state the results of experiments simply in words and avoid the abstractions and math?

Not all of science must be described in quantitative terms. That the sky appears blue in late morning and red toward evening is a wonderful bit of science that we can appreciate without math. So too, the life cycles of butterflies, the "V" formation of a flying flock of birds, the chambered structure of the heart, the change of the seasons. But much of science is quantitative, and here mathematics plays a crucial role.

Mathematics is useful for science for several reasons: mathematics expresses the underlying assumption that nature is logical and quantitative; mathematics provides a way to *generalize* the results of a small number of particular experiments and thus to make predictions about new experiments; mathematics is usually briefer and more convenient than words; and mathematics provides a way to proceed logically from one result to a second result, equivalent to the first but perhaps in a more useful form. Let's discuss each of these issues in turn.

Science is based on the premise that nature is logical, that every effect has a cause. If wheelbarrows could suddenly lift off the ground and start floating, then there would be no point in searching for the laws of nature. Scientists must have faith that all things in nature have a rational and logical explanation, even if every explanation has not yet been found. Furthermore, the logical workings of nature are quantitative: the square of the orbital period of each planet is proportional to the cube of its distance from the sun; any object dropped to the ground increases its speed by 32 feet per second every second; and so on. Mathematics, by its nature, is logical and quantitative. Thus mathematics is an ideal language for science.

Mathematics provides a way to generalize the results of a small number of experiments. In the experiments with the rising and falling masses, for example, we found that a mass of 5 units rose a height of 3 units while a mass of 3 units fell a height of 5 units. In a second experiment, we found that a mass of 2 units rose a height of 2 units while a mass of 1 unit fell 4 units. We could, of course, do many more experiments with different masses and heights, and we could tabulate all the results in a great book, but unless we found a general pattern for our results, we could never predict what would happen in any experiment we hadn't already done. Once we have guessed a general pattern, we can use mathematics to express that pattern. For the rising and falling masses, the pattern we recognize after a small number of experiments is: "the amount of rising mass multiplied by the vertical distance it rises equals the amount of falling mass multiplied by the vertical distance it falls." This is a general statement and refers to *any* pair of masses and heights, not just the ones tabulated in our great book of results. It can be used to predict results of experiments not yet performed. Similarly, Eq. (I–1) generalizes the results found by the mother in accounting for her child's blocks. She can use that equation to make predictions about specific situations she has not yet encountered.

A closely related feature of mathematics is that it provides a shorthand way of expressing results. For example, once we have discovered the general pattern for rising and falling masses, we might get tired of having to write out the words "the amount of rising mass multiplied by the vertical distance it rises equals the amount of falling mass multiplied by the vertical distance it falls." A more compact description is Eq. (I–3a):

$$m_1 \times (h_{1i} - h_{1f}) = m_2 \times (h_{2f} - h_{2i}).$$

Once we have invested a little time to learn what these symbols stand for, this equation is more convenient to use than the longer statement in words, although the two are equivalent.

Finally, mathematics provides a way to proceed logically from one result to another equivalent result. The usefulness of this feature is subtle, but extremely important. Let's again use the example of the rising and falling masses. Once we have expressed the general pattern of results in the mathematical form of Eq. (I–3a), we can then use the rules of mathematics to manipulate that equation into other, logically equivalent forms. For example, without doing any further experiments, we can go from Eq. (I–3a) to Eq. (I–3b) simply by adding and subtracting and rearranging terms:

$$m_1 \times h_{1i} + m_2 \times h_{2i} = m_1 \times h_{1f} + m_2 \times h_{2f}.$$

Equation (I–3b) is mathematically and logically equivalent to the previous equation, Eq. (I–3a). Yet its different form suggests new things to us. In particular, Eq. (I–3b) has the form of a conservation law: All the initial quantities are on one side of the equation and all the final quantities are on the other side. An initial quantity equals a final quantity; something is conserved. We have discovered a *conservation law*. And the existence of this conservation law might lead us to search for other conservation laws. There is something wonderful, almost magical, about what has happened here. All the information of Eq. (I–3b) is in fact in Eq. (I–3a), since the two are logically equivalent. But Eq. (I–3a) is not in the form of a conservation law and thus does not immediately announce its significance to us. In much of science, the equivalent equations are far more complex than Eq. (I–3a) and Eq. (I–3b), and the number of intermediate steps connecting them far greater. In such situations, the meaning and significance of the initial equation is even less apparent, and mathematics becomes much more essential in leading us to a more meaningful result.

You might ask yourself how you could possibly deduce a conservation law by rearranging the *words* of the statement "the amount of rising mass multiplied by the vertical distance it rises equals the amount of falling mass multiplied by the vertical distance it falls." Mathematics is the language of science.

C. KINETIC ENERGY

1. Energy of Motion

Suppose that a mass attached to the end of a string is set in motion, as shown in Fig. I–4. The mass comes to rest at the top of its swing, then falls back in the other direction, gains speed, reaches maximum speed at the bottom of its swing, rises on the other side, and again comes to rest at the top of its swing. Then the process repeats. If there were no friction and no outside disturbance, the pendulum would keep swinging forever.

If we make the string very light, the pendulum bob is essentially the only mass in the system. Since the bob changes height during its swing, its gravitational energy changes. Can we square this result with the conservation of energy? Possibly so, but only if there is a second form of energy, one associated with speed, and only if we can find a rule for measuring the new energy

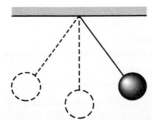

Figure I–4 A swinging pendulum.

that leads to a conservation law. We will call this energy of motion kinetic energy and denote it by the symbol E_K.

The pendulum has zero speed and therefore zero kinetic energy at the top of its swing, at maximum height, when its gravitational energy is largest; it has maximum speed and therefore maximum kinetic energy at the bottom of its swing, when its gravitational energy is smallest. Total energy will be conserved if the gain of kinetic energy exactly equals the loss of gravitational energy. Then there will be no net change in total energy as the pendulum swings back and forth. We now must see if there is a rule for counting kinetic energy such that the total energy of a system, $E_G + E_K$, stays constant. This is a quantitative question.

2. The Rule for Measuring Kinetic Energy

In searching for a rule for kinetic energy, we will build on our knowledge of gravitational energy. First we need to review the meaning of speed and to introduce the concept of acceleration.

Speed is the rate of change of distance. A car that travels 60 miles in an hour has an average speed of 60 miles per hour. The general relation between distance, average speed, and elapsed time is

(increase in distance) = (average speed) × (elapsed time).

If the car is moving in reverse, it is decreasing its distance, and the average speed is negative.

Acceleration is the rate of change of speed, just as speed is the rate of change of distance. If we are traveling in our car at 20 miles per hour and then increase our speed to 30 miles per hour, we have accelerated. We have changed our speed. A car that increases its speed from 20 miles per hour to 30 miles per hour in an hour has an average acceleration of 10 miles per hour per hour, since its speed has increased by 10 miles per hour over the course of an hour. The two "pers" in acceleration refer to a change in speed (distance per time) per interval of time. The definition of acceleration may be expressed as

(increase in speed) = (average acceleration) × (elapsed time).

If a moving object *decreases* its speed, the acceleration is negative. When a moving object changes direction, that change also constitutes an acceleration, but here we will be concerned only with changes in speed.

Let us now return to kinetic and gravitational energy. Suppose we drop an object to the ground, with no strings attached. It is experimentally observed that, in the absence of air resistance, all objects fall to the ground with the same acceleration, 32 feet per second per second. This important fact of nature was first recorded by the Italian physicist Galileo Galilei (1564–1642). If the speed of our falling object begins at zero, then after 1 second the speed is 32 feet per second, after 2 seconds the speed is 32 feet per second + 32 feet per second = 64 feet per second; after 3 seconds the speed is 64 feet per second + 32 feet per second = 96 feet per second, and so on. If the speed begins at 20 feet per second, then after 1 second the speed is 20 feet per second + 32 feet per second = 52 feet per second; after 2 seconds the speed is 52 feet per second + 32 feet per second = 84 feet per second, and so on. Each second the falling object picks up another 32 feet per second in speed. It falls faster and faster.

Not all accelerations are 32 feet per second per second, and not all accelerations are constant in time. When we drive our car, for instance, we can vary the acceleration by changing how we push the gas pedal. But any object freely falling to the earth experiences a universal acceleration of 32 feet per second per second. This terrestrial acceleration is related to the strength of the earth's gravity. An object dropped on the moon, which has smaller gravity than the earth's, falls with a smaller acceleration (5 feet per second per second), and an object dropped on Jupiter falls with a larger one (81 feet per second per second).

To continue our pursuit of kinetic energy, let's study the motion of an object dropped to the ground. So that we don't have to keep writing out "32 feet per second per second" for the acceleration, we will denote this number by g, which is called the gravitational acceleration. We will also denote the speed of the falling object by v, time by t, and vertical height by h. Suppose that at some initial time t_i the object has a vertical height h_i and a speed v_i, and at a later time t_f it has a vertical height h_f and a speed v_f. The situation is shown in Fig. I–5. We want to see how much the speed increases over this interval of time. Use the definition of acceleration. The increase in speed is

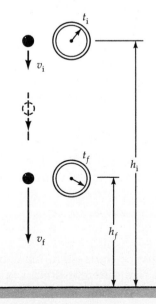

Figure I–5 A falling ball. At the initial time t_i the ball has speed v_i and height h_i. At a later time t_f the ball has speed v_f and height h_f.

$v_f - v_i$. This equals the average acceleration, g, multiplied by the elapsed time, $t_f - t_i$:

$$v_f - v_i = g(t_f - t_i). \tag{I–5a}$$

(Since g is constant, we don't have to worry about averaging it over the fall of the object.)

Notice that we left out the multiplication sign in Eq. (I–5a) and just wrote $g(t_f - t_i)$ instead of $g \times (t_f - t_i)$. This is a shorthand way of writing equations. Whenever two symbols are put side by side, they should be multiplied, and we will use this shorthand from now on.

What is the vertical distance $(h_i - h_f)$ the object falls during the time period $(t_f - t_i)$? The increase in distance equals the average speed multiplied by the elapsed time. In our case, the speed is continuously increasing, starting at v_i and ending at v_f. For a constant acceleration, as we have here, the average speed is $(v_i + v_f)/2$. Thus, the vertical distance the object falls during this time is

$$h_i - h_f = \frac{v_i + v_f}{2}(t_f - t_i). \tag{I–5b}$$

From Eq. (I–5a), we know $t_f - t_i = (v_f - v_i)/g$. Substituting this into Eq. (I–5b), we obtain

$$h_i - h_f = \frac{(v_i + v_f)(v_f - v_i)}{2g} = \frac{v_iv_f - v_i^2 + v_f^2 - v_fv_i}{2g} = \frac{v_f^2 - v_i^2}{2g}. \tag{I–5c}$$

Equation (I–5c) is beginning to look interesting. It relates a change in height, which characterizes a change in gravitational energy, to a change in speed, which characterizes a change in kinetic energy. Notice that Eq. (I–4), the formula for gravitational energy, involves a constant multiplied by mass multiplied by height. Using the equation $E_G = Cmh$ as a hint, multiply both sides of Eq. (I–5c) by mg, to obtain

$$mg(h_i - h_f) = \tfrac{1}{2}m(v_f^2 - v_i^2). \tag{I–6a}$$

Now rearrange Eq. (I–6a) by putting all the quantities at the initial time on one side and at the final time on the other:

$$mgh_i + \tfrac{1}{2}mv_i^2 = mgh_f + \tfrac{1}{2}mv_f^2. \tag{I–6b}$$

Aha! This is a conservation law. A certain quantity of the system, namely $mgh + mv^2/2$, remains the same when evaluated at two different times, even though the system has changed. If $mgh + mv^2/2 = 56$ units of energy at time t_i, for example, then $mgh + mv^2/2 = 56$ units of energy at the later time t_f, even though h and v have changed. Using only logic and algebra, we have discovered a new conservation law. We didn't have to assume that there was a total conserved quantity involving both gravity and motion. We deduced it.

Now, to identify the quantitative expression for kinetic energy, an energy associated with speed. We know that the full conservation law should say that $E_G + E_K$ is the same at an initial and final time. The first term on each side of Eq. (I–6b), mgh, has nothing to do with speed, while the second one, $mv^2/2$, does. Indeed, the first term looks like gravitational energy. Evidently, the second term is the expression for kinetic energy:

$$E_K = \tfrac{1}{2}mv^2. \tag{I–7}$$

Our new conservation law not only tells us how to measure kinetic energy. It also provides a natural value for the previously unspecified constant C in the formula for gravitational energy, Eq. (I–4). If we let $C = g$, then the formula for gravitational energy becomes

$$E_G = mgh, \tag{I–8}$$

which looks exactly like the first term on each side of Eq. (I–6b). Remember that for systems with only gravitational energy, C could have been any constant. By now requiring that $C = g$, we have found an expression for gravitational energy that holds both for systems with only gravitational energy and for systems with both gravitational and kinetic energy.

Equation (I–6b) can now be written as

$$E_{Gi} + E_{Ki} = E_{Gf} + E_{Kf}, \tag{I–9a}$$

where $E_{Gi} = mgh_i$, $E_{Ki} = \tfrac{1}{2}mv_i^2$, and so on. The sum of gravitational energy and kinetic energy in an isolated system, with no other energies involved, is constant; its value at an initial time equals its value at any later time. With the rule for counting kinetic energy, Eq. (I–7), the total energy of a mass is conserved. Although the gravitational energy and the kinetic energy of a mass may individually change, their sum does not. Moreover, just as for gravitational energy alone, Eq. (I–9a) can be applied to a collection of masses in a system. It then says that the total gravitational plus kinetic energy, summed over all the masses of the system, is constant. If there are two masses in the system, for example, labeled by 1 and 2, then Eq. (I–9a) becomes

$$E_{G1i} + E_{G2i} + E_{K1i} + E_{K2i} = E_{G1f} + E_{G2f} + E_{K1f} + E_{K2f}. \tag{I–9b}$$

When kinetic energy is not involved in the system, as in the initial and final configurations shown in Fig. I–1, that is, $E_{K1i} = 0$, $E_{K1f} = 0$, $E_{K2i} = 0$, $E_{K2f} = 0$, then Eq. (I–9b) simply says that the total gravitational energy is conserved.

We derived Eq. (I–6) only for a freely falling mass. What happens if the mass if not freely falling—for example, if it is being slowly lowered on a string? In that case, the mass does not constitute an isolated system. Outside forces are acting on it, and these forces generally add or subtract energy from the mass. In such a case, Eq. (I–6) does not apply. Remember that energy is conserved only in isolated systems, except in special cases where the outside forces do not change the energy of the system.

Problem I–2: Loopty Loop

A ball starts at rest at point A, 3 inches high, on a loopty loop, as shown in Fig. I–6. Assume that the ball can roll on the loop without any friction. The ball begins rolling down the loopty loop, following the arrows, until it comes to rest at point B, after which time it will start to roll back down the hill. How high is point B?

Solution: If we assume the ball has no friction with the loop, then the ball is an isolated system. Apply the law of conservation of energy, Eqs.(I–7), (I–8), and (I–9). The total energy must be the same at points A and B (and at any other point). At points A and B, the ball is at rest, so the kinetic energy at these two points is zero. The law of conservation of energy then reads $E_{GA} = E_{GB}$, or $mgh_A = mgh_B$, where we use the symbol h for height, as usual, and the subscripts A and B refer to positions A and B, respectively. The factor mg is common to both sides and may be canceled, leading to $h_A = h_B$. Thus, the height at B is equal to the height at A, 3 inches.

Figure I–6 A ball rolling along a frictionless surface. The ball starts at A, 3 inches above the ground, and comes to rest at B, before rolling backward.

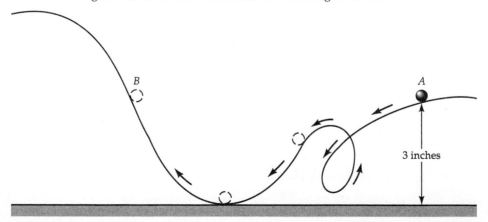

Problem I–3: The Speed of a Pendulum

Consider the pendulum shown in Fig. I–7. The pendulum bob is not an isolated system because the string exerts a force on it. However, a pendulum
is a special case in which the outside force of the string does not change
the energy of the bob. Thus, we may apply the law of the conservation of
energy to the bob by itself. (a) If the top of the bob's swing is 8 inches
above the bottom of its swing, what is the bob's speed at the bottom? This
speed is the maximum speed of the pendulum. (b) If the maximum height
of the pendulum bob is H, what is its speed v at any height h? (c) If a pendulum bob has a speed of 25 feet per second at the bottom of its swing,
how high is the bob at the top of its swing?

Solution: (a) Let h_i and v_i be the height and speed of the pendulum bob at
the top of its swing, and h_f and v_f be the height and speed at the bottom.
We are given that $h_i - h_f = 8$ inches, and we know that $v_i = 0$, since the
pendulum is at rest at the top of its swing. We are asked to find v_f. Apply
Eq. (I–9a). First, put it into the form

$$E_{Kf} = E_{Gi} - E_{Gf} + E_{Ki},$$

so that the unknown is on one side and the known on the other. Now,
$E_{Ki} = mv_i^2/2 = 0$ and

$$E_{Gi} - E_{Gf} = mg(h_i - h_f),$$

so we obtain

$$\frac{mv_f^2}{2} = mg(h_i - h_f).$$

Figure I–7 A swinging pendulum, which comes to rest at a height $H = 8$ inches above
its lowest point.

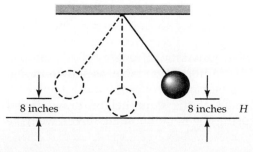

8 inches   8 inches H

Multiplying both sides by $2/m$ and taking the square root to solve for v_f, we arrive at

$$v_f = [2g(h_i - h_f)]^{1/2}$$
$$= (2 \times 32 \text{ feet per second per second} \times 0.67 \text{ feet})^{1/2}$$
$$= 6.5 \text{ feet per second.}$$

Note that in the last step, we had to convert inches to feet, 8 inches = 0.67 feet, since we expressed g in feet per second per second. The units must be the same throughout an equation; inches and feet cannot be mixed together. Our pendulum has a maximum speed of 6.5 feet per second. We can apply exactly the same method to find the speed of the ball at the bottom of the loop in Fig. I–6: $v = (2 \times 32 \text{ feet per second per second} \times 0.25 \text{ feet})^{1/2} = 4$ feet per second.

(b) Use the formula we just derived, $v_f = [2g(h_i - h_f)]^{1/2}$, and let $v_f = v$, $h_i = H$, and $h_f = h$. We then obtain the general formula for the speed v of a pendulum at any height h, given that its maximum height is H:

$$v = [2g(H - h)]^{1/2}. \tag{I–10}$$

Notice that Eq. (I–10) automatically guarantees that the speed is zero when $h = H$, at maximum height. That equation also shows that the maximum speed is $v = (2gH)^{1/2}$, when $h = 0$, at minimum height. At any intermediate value of h, it gives the speed at that value of h.

(c) If Eq. (I–10) is squared and solved for H, we get

$$H = h + \frac{v^2}{2g}.$$

We are given that $v = 25$ feet per second when $h = 0$. Substituting in these values, $H = 0 + (25 \text{ feet per second})^2/64 \text{ feet per second per second} = 9.8$ feet.

---

3. Systems with Only Kinetic Energy

We used our knowledge of gravitational energy to find the rule for adding up kinetic energy, Eq. (I–7), but once having discovered that rule, we can apply it to situations that don't involve gravitational energy. For example, consider a collision between two balls on a horizontal table, as shown in Fig. I–8. The balls come in with certain speeds, have a collision, and go out with different speeds. Since both balls remain at the same height, gravitational energy plays no role. More formally, using the notation in Eq. (I–9b), we have that

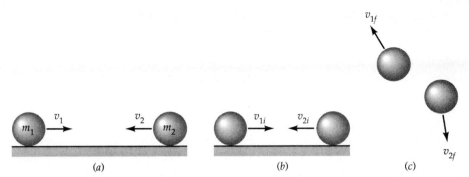

Figure I–8 (*a*) Two approaching balls, of masses m_1 and m_2 and speeds v_1 and v_2. (*b*) Two approaching balls, of speeds v_{1i} and v_{2i} before they collide. (*c*) After the collision, the two balls part, with final speeds v_{1f} and v_{2f}.

$$E_{G1i} = m_1 g h_i = m_1 g h_f = E_{G1f}.$$

The same goes for ball 2. Subtracting these equal quantities from both sides of Eq. (I–9b), we are left with

$$E_{K1i} + E_{K2i} = E_{K1f} + E_{K2f},$$

or, using the rule for kinetic energy, Eq. (I–7),

$$\tfrac{1}{2} m_1 v_{1i}^2 + \tfrac{1}{2} m_2 v_{2i}^2 = \tfrac{1}{2} m_1 v_{1f}^2 + \tfrac{1}{2} m_2 v_{2f}^2 \qquad (I\text{–}11)$$

Equation (I–11) doesn't involve gravitational energy at all. It relates the final speeds of two colliding balls to their initial speeds.

Problem I–4: Balls in Collision

Consider the collision between the two balls shown in Fig. I–8. Suppose that balls 1 and 2 have initial speeds of $v_{1i} = 7$ feet per second and $v_{2i} = 5$ feet per second. Ball 1 has a mass half as big as ball 2, $m_1 = m_2/2$. If ball 1 has a final speed $v_{1f} = \sqrt{65}$ feet per second, what is the final speed of ball 2?

Solution: Apply Eq. (I–11). Multiplying it by 2, dividing by m_2, and rearranging terms, we get

$$v_{2f}^2 = v_{2i}^2 + (m_1/m_2)(v_{1i}^2 - v_{1f}^2).$$

Substituting in the given values, this yields

$$v_{2f}^2 = 25(\text{feet per second})^2 + (\tfrac{1}{2})(49 - 65)(\text{feet per second})^2$$
$$= 17(\text{feet per second})^2.$$

Taking the square root of this gives $v_{2f} = \sqrt{17} = 4.1$ feet per second.

Historically, kinetic energy was the first type of energy for which a conservation law was known. The conservation of kinetic energy was considered in qualitative terms by the great French philosopher and mathematician René Descartes (1596–1650). Descartes is best known for his phrase *"Cogito ergo sum"* ("I think, therefore I am") and for his development of cartesian geometry. In his *Principles of Philosophy* (1644) Descartes presented the most exhaustive study of nature and philosophy since Aristotle's. This work proposed that all of nature can be understood in mechanistic terms. In his *Principles*, Descartes writes

> Now that we have examined the nature of motion, we come to consider its cause. . . . As for the first cause, it seems to me evident that it is nothing other than God, Who by His Almighty power created matter with motion and rest in its parts, and Who thereafter conserves in the universe by His ordinary operations as much of motion and of rest as He put in it in the first creation. . . . We also know that it is one of God's perfections not only to be immutable in His nature but also to act in a way which never changes . . . from which it follows that, since He set in motion in many different ways the parts of matter when He created them and since He maintained them with the same behavior and with the same laws as He laid upon them in their creation, He conserves continually in this matter an equal quantity of motion. [*Principles of Philosophy*, part II, Section 36]

Discussion Questions I–2:

In Descartes' view, what is the origin and basis of the conservation of kinetic energy? From this passage, do you think that Descartes experimentally verified the conservation of kinetic energy, or did he assume it?

The first quantitative statement of the conservation of kinetic energy was made in 1669, by the Dutch scientist and mathematician Christiaan Huygens (1629–1695). Huygens arrived at his result not by doing experiments with colliding balls, but rather by mathematical calculations and reasoning not unlike our path to Eq. (I–11).

D. UNITS OF LENGTH, MASS, WEIGHT, AND ENERGY

We have already solved several physical problems without having to specify the units in which mass and energy are measured. But assigning units will prove useful. In the example of the child's blocks in section A, the quantity being counted was simply the number of blocks, and so the unit was "one block." However, mass and energy do not come in blocks, so we will have to invent units. There is nothing sacred about units; they serve only as a convenience and a standard for communication. People know what you are talking about when you say the temperature is 40 degrees centigrade. If you say the temperature is 4 degrees Zanerbach, they don't.

We will base our units on the International System of Units, the centerpiece of which is the meter. In 1791, the French Academy of Sciences defined the meter as one ten-millionth the distance from the earth's equator to the North Pole. The meter is equal to about 39.37 inches, or 3.28 feet. Equivalently, 1 foot equals 0.305 meters. A centimeter is one hundredth of a meter. We can express the earth's gravitational acceleration in terms of meters instead of feet: g = 32 feet per second per second = 9.8 meters per second per second. From now on, we will measure lengths in meters.

A common unit of mass is the gram, which was originally defined to be the mass of one cubic centimeter of water. A penny has a mass of about 2.5 grams. One thousand grams is called a kilogram. An average person has a mass of about 70,000 grams, or 70 kilograms. Notice that the meter and the kilogram are based on natural terrestrial phenomena: the size of the earth and the density of water. So there is some rationality behind their definitions. (If we had nine fingers instead of ten, the meter might be defined as one nine-nillionth the quarter length of the earth's circumference, and so on.) The meter and the kilogram are the standard units of length and mass in the International System of Units.

Mass is the quantity of matter. It is not the same as weight. When you put an object on a scale to weigh it, you are measuring both the quantity of matter in that object and also how strongly the earth is pulling it down against the scale. A kilogram of matter would weigh much less on the moon than on the earth, since the moon's gravitational pull is much less than the earth's. (Despite their gear, the Apollo astronauts could jump 2 meters high on the moon.)

Weight is the product of mass and gravitational acceleration, mg. The familiar unit of weight is the pound. A mass of 1 kilogram on the earth, where g = 9.8 meters per second per second, has a weight of 2.2 pounds. Taking the reciprocal of this, we find that it takes about 0.454 kilograms to weigh a pound on earth. On the moon, where the gravitational acceleration is only 1.6 meters per second per second, 0.454 kilograms would weigh a sixth of a pound. An object's weight thus depends on where it is weighed. However, its mass does

TABLE I–1: Units

Length	Mass	Weight	Energy
1 meter	1 kilogram	1 pound	1 joule
100 centimeters 0.001 kilometers	1000 grams	0.454 kilograms $\times$ 9.8 meters/s/s	1 kilogram $\times$ (1 meter/s)^2
3.28 feet			0.24 calories

not. A kilogram is the same quantity of matter whether it sits in your living room or on Pluto.

Having defined the unit of mass, we can define the unit of energy. Since kinetic energy is mass multiplied by speed squared, energy will have units of mass multiplied by speed squared. For defining the unit of energy, we will measure mass in units of kilograms and speed in units of meters per second. Then a unit of energy will be a kilogram $\times$ (meter/second)^2. As this is a somewhat unwieldy combination of units, it has been given another name: a joule. One joule of energy equals 1 kilogram $\times$ (meter/second)^2. The joule is a common unit of energy. If all masses are put in kilograms and all speeds in meters per second, then the resulting energy will be in joules.

Gravitational energy must have the same units as kinetic energy, since both are energies. By analogy with our discussion for kinetic energy, if all masses are put in kilograms, all accelerations in meters per second per second, and all lengths in meters, the resulting energy will be in joules.

The units and conversions between units are summarized in Table I–1. Here, the quantities in each column are equal to each other, e.g., 1 meter equals 100 centimeters.

Problem I–5: Units of Energy

(a) A mass of 3 kilograms is traveling at a speed of 17 meters per second. How much kinetic energy does it have? (b) How much energy does a 1-pound edition of Charles Darwin's *Origin of the Species* gain when lifted a distance of 5 feet? What about a 2-pound stack of the *National Enquirer*? (c) How much energy does it take to lift a 10-ton elephant 0.5 meters?

Solution: (a) $E_K = mv^2/2 = 3$ kilograms $\times$ (17 meters/sec)$^2/2 = 434$ joules.
(b)

$$E_{Gf} - E_{Gi} = mg(h_f - h_i).$$

To express the answer in joules, we must put all masses in kilograms, all accelerations in meters per second per second, and all lengths in meters. One pound corresponds to 0.454 kilograms on the earth. The gravitational acceleration $g = 9.8$ meters per second per second. The height increase is $h_f - h_i = 5$ feet/(3.28 feet/meter) = 1.52 meters. Thus, the increase in energy is

$$E_{Gf} - E_{Gi} = 0.454 \text{ kg} \times 9.8 \text{ m/s/s} \times 1.52 \text{ m} = 6.76 \text{ joules.}$$

Here, for the units of mass, length, and time, we have used the abbreviations that kg stands for kilogram, m stands for meter, and s stands for second.

(c) There are 2000 pounds in a ton, so $m = 10$ tons $= 10$ tons $\times$ 2000 pounds/ton $\times$ 0.454 kilograms/pound = 9080 kilograms. Thus

$$E_G = 9080 \text{ kg} \times 9.8 \text{ m/s/s} \times 0.5 \text{ m} = 44{,}492 \text{ joules.}$$

Although we have used the subscripts K and G to indicate the form of energy, all energy is measured in the same units. An energy of 13 joules is an energy of 13 joules, no matter what form it's in.

E. CONSERVATION LAWS AND HUMAN FREEDOM

The above examples with pulleys and masses and pendulums and colliding balls illustrate the use and authority of the conservation of energy. Although energy may not be handled or even seen, it can be quantified. The law of the conservation of energy gives us the power to predict the workings of nature, and such power provides something of certainty and control in the world.

The philosophical and psychological meaning of a conservation law is nowhere better illustrated than in the atomistic theory of the ancient Greeks. Originated by Democritus in the late fifth century B.C., this theory proposed that the universe was made of atoms and empty space. Atoms (from the Greek word *atomos*, meaning "indivisible") were taken to be indestructible microscopic bodies that made up all substances. Atoms differed in properties—for example, some were hard and some were soft, some smooth and some thorny—and these differences explained the variation in substances throughout the universe. Atoms could be neither created nor destroyed. They were conserved, like the child's 28 blocks. Atoms were the constants in a changing world and thus gave some measure of psychological comfort and security. The Greek atomistic theory provided an explanation for everything from the nature of wind, to why fish have scales, why light but not rain passes through a horn, why corpses smell bad and saffron good, and even the origin of the cosmos. The theory became a world view.

The atomistic world view also permeates the classic poem *De Rerum Natura* (*The Nature of Things*), written by the Roman Lucretius (ca. 95–ca. 55 B.C.). Lucretius begins his poem by noting that people have two fears: helplessness under the power of the gods, and death.

> When man's life lay for all to see foully grovelling upon the ground, crushed beneath the weight of Superstition, which displayed her head from the regions of heaven, lowering over mortals with horrible aspect. (I, 62–66) . . . You will yourself some day or other seek to fall away from us, overborne by the terrific utterances of the priests. Yes indeed, for how many dreams can they [the priests] even now invent for you, enough to upset the principles of life and to confound all your fortunes with fear! And with reason; for if men saw that a limit has been set to tribulation, somehow they would have strength to defy the superstitions and threatenings of the priests; but, as it is, there is no way of resistance and no power, because everlasting punishment is to be feared after death. For there is ignorance what is the nature of the soul, whether it be born or on the contrary find its way into men at birth, and whether it perish together with us when broken up by death, or whether it visit the gloom of Orcus and his vasty chasms. (I, 102–117)

For Lucretius, ignorance of the workings of nature and of the fate of our souls are the principal causes of our fears. Lucretius then proposes a remedy: knowledge of the laws of nature and, in particular, knowledge that the fundamental substances of nature cannot be created or destroyed.

> This terror of mind therefore and this gloom must be dispelled, not by the sun's rays or the bright shafts of day, but by the aspect and law of nature. The first principle of our study we will derive from this, that no thing is ever by divine power produced from nothing. For assuredly a dread holds all mortals thus in bond, because they behold many things happening in heaven and earth whose causes they can by no means see, and they think them to be done by divine power. For which reasons, when we shall perceive that nothing can be created from nothing, then we shall at once more correctly understand from that principle what we are seeking, both the source from which each thing can be made and the manner in which everything is done without the working of the gods. (I, 146–158)

Finally, Lucretius discusses the basis of the conservation of things, the indestructible atoms, which he calls "first beginnings of things":

> Furthermore, bodies are partly the first-beginnings of things, partly those which are formed by union of the first-beginnings. But those which are the first-beginnings of things no power can quench: they conquer after

all by their solid body. (I, 483–487) . . . there are such things as consist of body solid and everlasting, which we teach to be seeds of things and their first-beginnings, out of which now all the sum of things has been built up. (I, 498–502)

Having argued that all substances are composed of indestructible atoms and thus free from the vagaries of the gods, Lucretius uses the atomistic theory to advocate the mortality of the soul and thus the impossibility of punishment after death.

> This same reasoning teaches that the nature of mind and spirit is bodily; for when it is seen to drive forward the limbs, to arouse the body from sleep, to change countenance, to guide and steer the whole man, and we see that none of these things can be done without touch, and further that there is no touch without body, must we not confess that mind and spirit have a bodily nature? (III, 161–169) . . . Now I shall go on to explain to you, of what kind of body this mind is, and of what it is formed. First I say that it is exceedingly delicate and formed of exceedingly minute particles [the first beginnings, or atoms]. (III, 177–181) . . . when for example I speak of spirit, showing it to be mortal, believe me to speak also of mind, inasmuch as it is one thing and a combined nature. First of all, since I have shown it to be delicate and composed of minute particles and elements much smaller than the flowing liquid of water or cloud or smoke . . . therefore, since, when vessels are shattered, you perceive the water flowing out on all sides and the liquid dispersing, and since mist and smoke disperse abroad into the air, believe that the spirit also is spread abroad and passes away far more quickly, and is more speedily dissolved into its first bodies, as soon as it has departed withdrawn from the limbs of a man. (III, 422–440) . . . Therefore death is nothing to us, it matters not one jot, since the nature of the mind is understood to be mortal. (III, 830–833) . . For, if by chance anyone is to have misery and pain in the future, he must himself also exist then in that time to be miserable. Since death takes away this possibility, and forbids him to exist for whom these inconveniences may be gathered together, we may be sure that there is nothing to be feared after death. (III, 862–869) [*De Rerum Natura* by Lucretius, translated and edited by W. H. D. Rouse and M. F. Smith, Loeb Classical Library (Harvard University Press: Cambridge, Massachusetts, 1982)].

Discussion Questions I–3

How does it limit the power of the gods that things cannot be created or destroyed? How does the indestructibility of atoms give human beings more control over their own fate? More peace of mind?

Discussion Questions I–4

Why did Lucretius believe in something that couldn't be seen, namely the atoms? Why do modern scientists believe in things they cannot see? How do you think modern scientists would react to phenomena that do not appear to conserve energy?

Discussion Questions I–5

Are the two human fears that Lucretius begins with—fear of the power of the gods and fear of death—still present today? Are they present in some other form? If so, would a conservation law alleviate these modern fears?

Discussion Questions I–6

Are there concepts in art or history or law that correspond to conservation laws, that is, constants of form or human nature or action?

F. HEAT ENERGY

1. Friction and a New Form of Energy

Consider again a swinging pendulum. We have assumed that it is an isolated system, so that its total energy is conserved. In such a situation, the pendulum will keep swinging forever. But we know this doesn't happen. Eventually, a pendulum that is not rewound or lifted will slow down and stop. Why? And where does the energy go?

Any real pendulum is not an isolated system; it is surrounded by air, and the molecules of air bump into the swinging pendulum. On average, the molecules move faster after bumping into the pendulum. These collisions are a form of friction, which removes energy from the pendulum and transfers it to the molecules of air. Even though the individual air molecules cannot be seen, they can absorb energy from the pendulum. Friction is also present in the joint that holds the pendulum and in the collisions between the air molecules and the walls of the room, but it is sufficient for our purposes (and a good approximation) to ignore these other forms of friction and consider only the friction between the pendulum and the air.

If the pendulum and its surrounding air are placed in a box, then the total system of pendulum plus air *is* isolated. Then the total energy of that system is conserved. A sensitive thermometer placed in the box would show that the temperature of the air rises as the pendulum slows down. In other words, the energy of the pendulum goes into heating the air. With the right rules for counting, the sum of the gravitational and kinetic energy of the pendulum and the heat energy of the air remains constant; when the first decreases, the second increases. The equation of the conservation of energy now reads

$$E_G + E_K + E_H = \text{constant}, \tag{I–12}$$

where E_H is the heat energy.

2. The Nature of Heat

Heat energy is a special form of kinetic energy in which many masses move randomly in all directions instead of together in the same direction. Figure I–9 illustrates the difference between the random motions that characterize heat energy and the ordered motions that characterize bulk kinetic energy. Heat energy does not cause an *overall* motion of the group of masses. Hot chicken soup in a pot, for example, doesn't leap out of the pot. Individual molecules move but are constantly changing directions, so that there is no net motion over time. By contrast, bulk kinetic energy is associated with an overall motion because the masses all continue moving in the same direction. In a swinging pendulum, for example, the molecules of the bob move together in bulk, in the same direction. From one moment to the next, the pendulum bob changes position. We will call ordered, or bulk, kinetic energy simply kinetic energy, and we will call random kinetic energy heat energy.

It wasn't until the nineteenth century that the nature of heat was understood. Before this time, many people thought heat was a fluid, called caloric, that could be passed from one body to another. The fluid hypothesis appeals to common sense. When a hot body touches a cold one, the hot body cools and the cold body grows hotter, as if some substance were being passed from one to the other. The British mathematician Brook Taylor (1685–1731) performed a

Figure 1–9 (*a*) Particles moving in random directions, illustrating heat energy. (*b*) Particles all moving in the same direction.

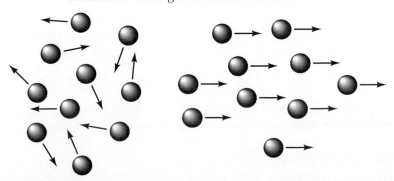

(*a*) Random motions   (*b*) Ordered motions

number of experiments in which he mixed hot and cold water together and was able to predict the resulting temperature by assuming that heat was a conserved "quantity." Summarizing his results, he wrote, "Having first observed where the thermometer stood in cold water, I found that its rising from that mark, or the expansion of the oil, was accurately proportional to the quantity of hot water in the mixture, that is to the degree of heat." (*Philosophical Transactions*, vol. 6, part II, p. 17) The Scottish scientist Joseph Black (1728–1799), a founder of modern chemistry and pioneer in the study of heat, referred to heat as "distributing" itself in a body and wrote that different bodies with the same temperature and same mass may nevertheless have "different quantities of the matter of heat." [*Lectures on the Elements of Chemistry*, portions of which are reprinted in *A Source Book in Physics*, ed. W. F. Magie (McGraw-Hill: New York, 1935)]

Although a number of scientists had hypothesized that heat was motion rather than substance, much of the credit for firmly establishing this view goes to Benjamin Thompson (1753–1814). At first a school teacher in Rumford (now Concord), New Hampshire, Thompson served in the American army at the beginning of the Revolutionary War but fled to England after the fall of Boston. Highly successful in European government and military circles, he became head of the Bavarian army and was named Count Rumford by the Bavarian duke. The multitalented Rumford recounted the following in an essay read before the Royal Society of London in 1798:

> It frequently happens that in the ordinary affairs and occupations of life, opportunities present themselves of contemplating some of the most curious operations of Nature; and very interesting philosophical experiments might often be made, almost without trouble or expense, by means of machinery contrived for the mere mechanical purpose of the arts and manufactures. . . .

> Being engaged, lately, in superintending the boring of cannon, in the workshops of the military arsenal at Munich, I was struck with the very considerable degree of Heat which a brass gun acquires, in a short time, in being bored; and with the still more intense Heat (much greater than that of boiling water as I found by experiment) of the metallic chips separated from it by the borer.

> The more I meditated on these phenomena, the more they appeared to me to be curious and interesting. A thorough investigation of them seemed even to bid fair to give a farther insight into the hidden nature of Heat; and to enable us to form some reasonable conjectures respecting the existence or non-existence of an igneous fluid: a subject on which the opinions of philosophers have, in all ages, been much divided. . . .

> From whence comes the Heat actually produced in the mechanical operation above mentioned? Is it furnished by the metallic chips which are separated by the borer from the solid mass of metal?

If this were the case, then, according to the modern doctrines of latent Heat, and of caloric, the capacity for Heat of the parts of the metal, so reduced to chips, ought not only to be changed, but the change undergone by them should be sufficiently great to account for all the Heat produced.

But no such change had taken place; for I found, upon taking equal quantities, by weight, of these chips, and of thin slips of the same block of metal separated by means of a fine saw, and putting them, at the same temperature (that of boiling water), into equal quantities of cold water, the portion of water into which the chips were put was not, to all appearance, heated either less or more than the other portion, in which the slips of metal were put . . . . From whence it is evident that the Heat produced [by boring the cannon] could not possibly have been furnished at the expense of the latent Heat of the metallic chips. (pp. 151–152)

In other words, Rumford found no change in the metal chips scraped out of the cannon. If those chips themselves had been the source of the heat, then they would be expected to have undergone some alteration after the production of so much heat. Count Rumford goes on to describe several experiments in which he measured the heat generated by boring cannons under various controlled conditions. Then he concludes

We have seen that a very considerable quantity of Heat may be excited in the Friction of two metallic surfaces, and given off in a constant stream or flux, in all directions, without interruption or intermission, and without any signs of diminution or exhaustion. From whence came the Heat which was continually given off in this manner, in the foregoing Experiments? . . .

Was it furnished by the air? This could not have been the case; for, in three of the Experiments, the machinery being kept immersed in water, the access of the air of the atmosphere was completely prevented.

Was it furnished by the water which surrounded the machinery? That this could not have been the case is evident: first, because this water was continually *receiving* Heat from the machinery, and could not, at the same time, be *giving to*, and *receiving Heat from*, the same body; and secondly, because there was not chemical decomposition of any part of this water.

And, in reasoning on this subject, we must not forget to consider that most remarkable circumstance, that the source of the Heat generated by friction in these Experiments appeared evidently to be *inexhaustible*.

It is hardly necessary to add that any thing which any *insulated* body, or system of bodies, can continue to furnish *without limitation* cannot possibly be *a material substance*: and it appears to me to be extremely difficult, if not quite impossible, to form any distinct idea of any thing capable of being excited and communicated in the manner the Heat was excited and communicated in these Experiments, except it be MOTION. (pp. 160–161) [*Collected Works* of Count Rumford, vol. II, essay IX; reprinted in *A Source Book in Physics*, op. cit.; page numbers refer to the latter book.]

The above passage shows an excellent combination of experiment and reasoning, characteristic of good science.

Discussion Questions I–7

Describe the logic by which Count Rumford concludes that heat cannot be a material substance. Do you think the logic is sound? How important were Rumford's own experiments to his conclusions?

Discussion Questions I–8

Do the above excerpts reveal anything about Rumford's personal qualities? If so, what?

Discussion Questions I–9

Do you think that fundamental scientific discoveries can still be made today by carefully observing the "ordinary affairs and occupations of life?" If not, why not?

3. Measuring Heat: Temperature

In general, a group of masses might have both bulk and random motions. To measure the heat energy of such a group, imagine catching up with it and moving alongside it, as if you were riding a bicycle alongside a traveling swarm of bees. Once you are traveling at the same speed as the hive, it appears to be at rest. From your new perspective, the bulk motion has gone. The remaining motions of individual bees as they buzz this way and that represent the heat energy. Suppose that the speeds of the masses from your new perspective are v_1, v_2, v_3, and so on. Suppose also, for simplicity, that all the masses are the same, m. Then the heat energy is

$$(\text{\# blocks seen}) + (\text{\# blocks under bed}) \tag{I–13a}$$

If there are N masses and N is a very big number, then it is convenient to define an average speed squared: $<v^2> = (v_1^2 + v_2^2 + v_3^2 + ...)/N$. In terms of this average, the heat energy is

$$E_H = \tfrac{1}{2}Nm < v^2 > = \tfrac{1}{2}M < v^2 >, \qquad\qquad \text{(I–13b)}$$

where $M = Nm$ is the total mass. Notice that heat energy involves the average *squared* speed, which is very different from the average speed. The average speed of molecules in an object at rest is zero. For an object at rest, the positive and negative speeds, corresponding to molecules going forward and backward, cancel each other out to give an average of zero. But the *squares* of those speeds are all positive and cannot cancel to zero.

Heat energy is measured with a thermometer, which is a device that responds to the *average* random motions of a group of molecules or other masses. When a thermometer is placed in a liquid, for example, the molecules of the liquid shake the molecules in the thermometer, and vice versa. This molecular dance causes some property of the thermometer, such as the height of its mercury, to change.

As we will show in the next chapter, if the molecules in the liquid have a higher *average* energy ($\tfrac{1}{2}m < v^2 >$) than the molecules in the thermometer, the molecules in the thermometer will gain energy and move about more rapidly. If the molecules in the liquid have a lower average energy, the reverse happens, and it is the molecules in the liquid that gain energy. Once the average energy of a molecule in the liquid and a molecule in the thermometer have equalized, the energy transfer will stop, the molecules in the thermometer will maintain their energy without further change, and the level of mercury in the thermometer will stop changing. At this point, the level of mercury measures something important about the liquid. It measures the average heat energy of a molecule in the liquid, that is, the energy per molecule. *This average molecular energy is called the temperature of the liquid.* The amount of liquid is irrelevant to its temperature. Place a thermometer in a bathtub and it will register the same temperature as when it is placed in a cup of water scooped from the same tub. Although there is much more water in the tub than in the cup, the *average* energy of a molecule in the tub and the cup is the same.

To be more precise, the temperature T is defined as

$$T = \frac{1}{2}\frac{< v^2 >}{c} = \frac{E_H}{Mc}, \qquad\qquad \text{(I–14)}$$

where E_H is the heat energy previously defined, $M = Nm$ is the total mass of the substance, and c is a constant called the specific heat. The ratio E_H/Mc is clearly proportional to the average energy per molecule. Double the number of molecules N for the same total heat energy E_H, and E_H/Mc drops by half; the average energy per molecule drops by half. On the other hand, temperature is independent of volume for a fixed total mass and heat energy. A fixed mass of gas that expands into a vacuum increases its volume, but its total mass, heat energy, and temperature do not change. The specific heat c, which we will discuss later, varies from one substance to another according to the variation in molecular properties of different substances, such as the variation in masses of molecules.

People think scientists are crazy because of all the notation and weird symbols they need to talk to each other. The symbol c for specific heat is no exception. In the late eighteenth century, some physicists decided to denote specific heat by the symbol c, for no obvious reason. We will continue that tradition.

Equation (I–14) also shows how the temperature of a fixed mass M changes when the heat energy changes. If T_i is the initial temperature for an initial heat energy E_{Hi} and T_f is the final temperature for a final heat energy E_{Hf}, then Eq. (I–14) says that $T_i = E_{Hi}/Mc$ and $T_f = E_{Hf}/Mc$. Subtracting the first of these equations from the second, we get

$$T_f - T_i = \frac{E_{Hf} - E_{Hi}}{Mc}. \qquad \text{(I–15a)}$$

From now on, to save space, we will use an abbreviation for the difference between final and initial values of a quantity. For example, the final temperature minus the initial temperature will be denoted by ΔT:

$$T_f - T_i = \Delta T.$$

A positive ΔT means that the final temperature is greater than the initial temperature, that is, the temperature has increased. A negative ΔT means that the final temperature is less than the initial, and the temperature has decreased. The same notation will be used for heat energy, $E_{Hf} - E_{Hi} = \Delta E_H$, and so on for other quantities. Using these abbreviations, we can write Eq. (I–15a) as

$$\Delta T = \frac{\Delta E_H}{Mc}. \qquad \text{(I–15b)}$$

4. Units of Heat

Historically, the units of heat were based on water, the universal liquid. Temperature is often measured on the Celsius scale, invented by the Swedish astronomer Anders Celsius in 1742. On this scale, water boils at 100 degrees C and freezes at 0 degrees C. (Centigrade is another name for the Celsius scale.) These two points fix the size of a Celsius degree and the zero mark.

Physicists generally use a different scale for temperature, the Kelvin scale, proposed in 1851 by the Scottish physicist William Thomson (Lord Kelvin). The Kelvin temperature scale is defined by $T_{Kelvin} = T_{Celsius} + 273$. Thus, water freezes at $T = 0 + 273 = 273$ on the Kelvin scale. The temperature T that appears in Eq. (I–14) is in Kelvin, for which zero degrees corresponds to zero heat energy. Zero degrees Kelvin is called absolute zero. It is the coldest that any substance can be. From Eq. (I–14), we see that absolute zero means that $<v^2> = 0$, which can be achieved only if every single molecule in the substance is motionless (aside from ordered motions). If even a single molecule is moving, then $<v^2>$ will be larger than zero, and the substance will have a temperature higher than absolute zero. To date, scientists have been able to cool substances to temperatures as low as 0.000001 degrees Kelvin, or 0.000001 degrees above absolute zero.

Note that since the Kelvin temperature is simply the Celsius temperature shifted by a constant amount, a *change* of 1 degree Kelvin corresponds to a change of 1 degree Celsius. When only *changes* of temperature, ΔT, are involved, the result is the same for Kelvin or Celsius and we will frequently use the more familiar Celsius.

A common unit of heat, called the calorie (named after the Latin word for heat, calor), is defined as the heat energy needed to raise the temperature of 1 gram of water by 1 degree Celsius. However, heat energy, like all other forms of energy, can also be measured in joules. From experiment, we have found that it takes 4.2 joules to raise the temperature of 1 gram of water by 1 degree Celsius, so

$$1 \text{ calorie} = 4.2 \text{ joules}. \qquad (I–16)$$

This odd relation between calories and joules is typical of situations where two systems of units were developed independently. The joule derives from the meter, which is based on the circumference of the earth, while the calorie derives from the thermal properties of water.

From Eq. (I–15b), we see that specific heat c has units of joules per kilogram per degree Celsius, since we can rewrite that equation as $c = \Delta E_H / M \Delta T$. If it takes 4.2 joules to raise the temperature of 1 gram of water by 1 degree Celsius, then it takes 4.2 joules $\times$ 1000 grams/kilogram = 4200 joules to raise the temperature of 1 kilogram of water by 1 degree. Setting $M = 1$ kilogram, $\Delta T = 1$ degree C, $\Delta E_H = 4200$ joules, we see that $c = 4200$ joules per kilogram per degree C for water. The specific heat of other substances is different. For lead, for example, $c = 126$ joules per kilogram per degree C. In other words, it takes 126 joules of heat energy to raise the temperature of lead by 1 degree.

Looking at Eq. (I–15), we see that the lower specific heat for lead means that the same quantity of heat energy in the same total mass will produce a higher temperature in lead than in water. This difference is mostly due to the fact that an atom of lead is much heavier than a molecule of water. Consequently, for the same total mass of lead and water, there are many fewer atoms of lead than molecules of water. If the same total energy is distributed among these particles, the *average* energy of a lead atom will thus be much larger than the *average* energy of a water molecule. Temperature measures average energy.

Problem I–6: Running off Calories

Running involves a lot of friction. Because of friction with the road and within the body itself, a runner has to replenish almost her full kinetic energy with every stride. In other words, in terms of energy requirements, a runner effectively comes to rest at the end of each stride and then must boost herself back up to speed. If a runner weighs 150 pounds, has a stride of 4 feet, and runs a mile in 8 minutes, how many calories does she burn up in a mile? Note: "Food calories" and physics calories are different. It takes 1000 physics calories to make 1 food calorie.

Solution: First convert to metric units. The runner has a mass of 150 pounds. One pound is equivalent to $1/2.2 = 0.45$ kilograms, so the runner has a mass of $150 \times 0.45 = 68$ kilograms. The runner's speed is 1 mile per 8 minutes, or 5280 feet per 8×60 seconds, or $5280/480 = 11$ feet per second. One foot is $1/3.28 = 0.30$ meters. Thus the speed in meters per second is 11 feet per second $\times$ 0.30 meters per foot $= 3.3$ meters per second. The kinetic energy of the runner at any moment is therefore

$$E_K = \tfrac{1}{2}mv^2 = \tfrac{1}{2}(68 \text{ kg}) \times (3.3 \text{ m/s})^2 = 370 \text{ joules.}$$

This amount of energy must be replaced at every stride. If each stride is 4 feet, there are $5280/4 = 1320$ strides taken in the mile. Therefore, the total energy consumed in a mile is $370 \times 1320 = 488{,}400$ joules, which equals 116,286 calories, or about 120 food calories. Not much for the effort!

5. Heat and the Conservation of Energy

Heat is one form of energy and is part of the law of conservation of energy, as given in Eq. (I–12). That law can be written as

$$E_{Gi} + E_{Ki} + E_{Hi} = E_{Gf} + E_{Kf} + E_{Hf},$$

where the subscripts i and f refer to initial and final values, as usual. This relation can be rearranged to say

$$E_{Hf} - E_{Hi} = E_{Gi} - E_{Gf} + E_{Ki} - E_{Kf}, \qquad \text{(I–17a)}$$

or

$$\Delta E_H = -\Delta E_G - \Delta E_K. \qquad \text{(I–17b)}$$

In words, Eq. (I–17b) says that a substance's *increase* in heat energy equals its *decrease* in gravitational and kinetic energy, which is just a statement that its total energy remains constant.

Problem I–7: Dropping Bricks into Puddles

A 5-kilogram brick is dropped from the top of the humanities building, 15 meters high, into a puddle containing 30 kilograms of water. If all the energy of the brick goes into heating the water, how much does the temperature of the water rise?

Solution: The initial gravitational energy of the brick is $E_{Gi} = mgh_i = 5$ kg $\times$ 9.8 m/s/s $\times$ 15 m = 735 joules. The final gravitational energy of the brick is zero, $E_{Gf} = 0$. The initial and final kinetic energies of the brick are also zero. Substituting these values into Eq. (I–17b) gives $\Delta E_H = 735$ joules. Substituting this into Eq. (I–15b), with $M = 30$ kg *and* $c = 4200$ joules per kilogram per degree, we obtain

$$\Delta I = \frac{735 \text{ joules}}{30 \text{ kg } 4200 \text{ j } / \text{kg} / \text{C}} = 0.0058 \text{ C}.$$

This is how much the water temperature rises.

Problem I–8: Heat Exchange

Consider two tanks of gas, one containing 1 kilogram of oxygen at a temperature of 100 degrees C and one containing 0.7 kilograms of helium at 150 degrees C. The specific heats of oxygen and helium are $c = 655$ joules per kilogram per degree and $c = 3150$ joules per kilogram per degree, respectively. If the two tanks are placed into thermal contact with each other, as in Fig. I–10, they can exchange heat but not gas. Because of the heat exchange, the two gases eventually come to the same temperature T. What is T?

Solution: Apply the conservation of energy to the total system, consisting of the two gases. Let the oxygen be labeled 1 and the helium 2. Only heat energy is involved in this problem; gravitational and kinetic energy play no role. Thus, the conservation of energy says that

$$E_{H1i} + E_{H2i} = E_{H1f} + E_{H2f},$$

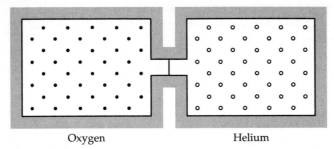

Oxygen   Helium

Figure I–10 Tanks of oxygen gas and hydrogen gas, separated by a plate that allows the exchange of heat between the two gases. Each small circle or dot indicates a gas molecule.

where E_{H1i} is the initial heat energy of the oxygen gas, and so on. Rearranging this, and using our Δ notation, we have

$$-\Delta E_{H1} = \Delta E_{H2}.$$

Substituting in $E_H = McT$, we have

$$- M_1 c_1 (T_{1f} - T_{1i}) = M_2 c_2 (T_{2f} - T_{2i}).$$

We are given that $M_1 = 1$ kg, $M_2 = 0.7$ kg, $c_1 = 655$ joules/kg/C, $c_2 = 3150$ joules/kg/C, $T_{1i} = 100$, $T_{2i} = 150$, and $T_{1f} = T_{2f} = T$. Substituting these values into our equation:

$$-1 \text{ kg} \times 655 \text{ j/kg/C} \times (T - 100) = 0.7 \text{ kg} \times 3150 \text{ j/kg/C} \times (T - 150),$$

or $65{,}500 - 655T = 2205T - 330{,}750$. This last equation gives $T = 138.5$ C.

Historically, heat played a crucial role in formulating the idea of the conservation of energy. Gravitational energy and kinetic energy are clearly related, since any dropped object picks up speed in a visible way. By contrast, the conversion between these forms of energy and heat energy is not so obvious. Heat consists of the random motions of microscopic particles and is not visible to the eye. Moreover, the sensation of heat is quite different from any human experience with gravitational or kinetic energy. The realization in the nineteenth century that heat was a form of energy, able to change into other forms of energy according to a strict accounting, gradually led scientists to the notion of the conservation of total energy as a general principle.

The German physician Julius Robert Mayer (1814–1878) is generally credited for first proposing an equivalence of all forms of energy, including heat, and a conservation of total energy. In his pioneering essay published in *Annalen der Chemie und Pharmacie* in 1842, Mayer writes

> Energies are causes: accordingly, we may in relation to them make full application of the principle—*causa aequat effectum* [cause equals effect]. If the cause c has the effect e, then $c = e$; if, in its turn, e is the cause of a second effect f, we have $e = f$, and so on: $c = e = f \ldots = c$. In a chain of causes and effects, a term or a part of a term can never, as plainly appears from the nature of an equation, become equal to nothing. The first property of all causes we call their *indestructibility*.

Mayer is saying here that any phenomenon, or cause, must have some effect. He goes on:

> If the given cause c has produced an effect e equal to itself, it has in that very act ceased to be: c has become e . . . Accordingly, since c becomes e, and e becomes f, we must regard these various magnitudes as different forms under which one and the same object makes its appearance. This capability of assuming various forms is the second essential property of all causes. Taking both properties together, we may say causes are (quantitatively) *indestructible* and (qualitatively) *convertible* objects. (pp. 197–198)

Here, Mayer is saying that phenomena (causes) may change into other phenomena but cannot simply disappear. He then summarizes and infers the necessary existence of another form of energy needed to conserve the total energy:

> In numberless cases we see motion cease without having caused another motion or the lifting of a weight; but energy once in existence cannot be annihilated, it can only change its form; and the question therefore arises, What other forms is energy, which we have become acquainted with as falling energy [gravitational energy] and motion, capable of assuming? . . . If, for example, we rub together two metal plates, we see motion disappear, and heat, on the other hand, make its appearance . . . Without the recognition of a causal connexion between motion and heat, it is just as difficult to explain the production of heat as it is to give any account of the motion that disappears. (p. 200)

> The falling of a weight is a real diminution of the bulk [gravitational energy] of the earth, and must therefore without doubt be related to the quantity of heat thereby developed; this quantity of heat must be proportional to the greatness of the weight and its distance from the ground. From this point of view we are very easily led to the equations between falling energy, motion, and heat, that have already been discussed. (p. 201) [*Annalen der Chemie und Pharmacie*, vol. 42, p. 233 (1842); translated by G. C. Foster in *Philosophical Magazine*, series 4, vol. 24, p. 271 (1862);

reprinted in *a Source Book in Physics*, op. cit.; page numbers refer to this last book. Note: In the German original, the word *Kraft*, meaning force, is used for energy.]

Discussion Questions I–10

What relationships, if any, do you see between Mayer's discussion of causes and Lucretius's analysis of phenomena in terms of indestructible atoms? From the above excerpt, do you think Mayer's argument for the conservation of energy derived more from philosophical belief or from quantitative observations and results?

Although Mayer first proposed the general law of the conservation of energy, it was the British physicist James Prescott Joule (1818–1889) who first put the law on firm footing, through his own experiments while in his twenties. The son of a wealthy brewer, Joule never worked for a living and built his laboratories at his own expense. He reported some of his results at a lecture titled "On Matter, Living Force, and Heat," given at St. Ann's Church Reading Room in Manchester on April 28, 1847. The following are excerpts from that lecture (note that Joule uses the words "living force" to denote kinetic energy):

From these facts it is obvious that the force [energy] expended in setting a body in motion is carried by the body itself, and exists with it and in it, throughout the whole course of its motion. This force [energy] possessed by moving bodies is termed by mechanical philosophers *vis viva*, or *living force* [kinetic energy]. The term may be deemed by some inappropriate, inasmuch as there is no life, properly speaking, in question; but it is *useful*, in order to distinguish the moving force [energy] from that which is stationary in its character, as the force of gravity. . . . The living force of bodies is regulated by their weight [mass] and by the velocity of their motion. (p. 80)

A body may be endowed with living force in several ways. It may receive it by the impact of another body. . . . A body may also be endowed with living force by means of the action of gravitation upon it through a certain distance. If I hold a ball at a certain height and drop it, it will have acquired when it arrives at the ground a degree of living force proportional to its weight and the height from which it has fallen. We see, then, that living force may be produced by the action of gravity through a given distance or space. We may therefore say that the former is of equal value, or *equivalent*, to the latter. (p. 81)

You will at once perceive that the living force of which we have been speaking is one of the most important qualities with which matter can be endowed, and, as such, that it would be absurd to suppose that it can be destroyed, or even lessened, without producing the equivalent of attraction through a given distance of which we have been speaking. You will therefore be surprised to hear that until very recently the universal opinion has been that living force could be absolutely and irrevocably destroyed at any one's option. Thus, when a weight falls to the ground, it has been generally supposed that its living force is absolutely annihilated, and that the labour which may have been expended in raising it to the elevation from which it fell has been entirely thrown away and wasted, without the production of any permanent effect whatever. We might reason, *a priori*, that such absolute destruction of living force cannot possibly take place, because it is manifestly absurd to suppose that the powers with which God has endowed matter can be destroyed any more than that they can be created by man's agency. (p. 82)

How comes it to pass that, though in almost all natural phenomena we witness the arrest of motion and the apparent destruction of living force, we find that no waste or loss of living force has actually occurred? Experiment has enabled us to answer these questions in a satisfactory manner; for it has shown that, wherever living force is *apparently* destroyed, an equivalent is produced which in process of time may be reconverted into living force. This equivalent is *heat*. Experiment has shown that wherever living force is apparently destroyed or absorbed, heat is produced. The most frequent way in which living force is thus converted into heat is by means of friction. . . . Heat, living force, and attraction through space [gravitational energy] are mutually convertible into one another. In these conversions nothing is ever lost. The same quantity of heat will always be converted into the same quantity of living force. (pp. 82–83)

Indeed the phenomena of nature, whether mechanical, chemical, or vital [biological], consist almost entirely in a continual conversion of attraction through space, living force, and heat into one another. Thus it is that order is maintained in the universe—nothing is deranged, nothing is lost, but the entire machinery, complicated as it is, works smoothly and harmoniously. And though, as in the awful vision of Ezekiel, "wheel may be in the middle of wheel," and everything may appear complicated and involved in the apparent confusion and intricacy of an almost endless variety of causes, effects, conversions, and arrangements, yet is the most perfect regularity preserved—the whole being governed by the sovereign will of God. (p. 85) [*The Scientific Papers of James Prescott Joule*, The Physical Society, London (1884), pp. 265–276; reprinted in *Kinetic*

Theory: Volume I, The Nature of Gases and Heat, ed. S. G. Brush (Pergamon Press: New York, 1965); pages refer to the last book.]

Discussion Questions I–11

Both Joule and Lucretius relate a conservation law to the power of God, but in very different ways. Discuss the differences. Was there any difference between the role of the gods in Lucretius' time and in Joule's?

Discussion Questions I–12

What is the meaning of Joule's reference to "order" and "regularity" in nature at the end of the above excerpt? Can you tell from his writing how he feels about these ideas and about the existence of a conservation law? Compare Joule's discussion and vocabulary here to the passage from Alexander Pope's poem *Windsor Forest* at the beginning of the chapter.

---

6. Heat, Work, and the First Law of Thermodynamics

Energy is the capacity to do work, and work always involves a change in energy. For example, lifting a book requires work because the gravitational energy of the book is being increased. Hitting a tennis ball requires work because the kinetic energy of the ball is being increased.

A famous law of physics, called the first law of thermodynamics and articulated in the nineteenth century, involves work and heat energy. The "thermo" refers to the heat, and the "dynamics" refers to the work. As we will see, the first law of thermodynamics is nothing more than the conservation of energy.

To arrive at the first law of thermodynamics, let us consider three systems, labeled 1, 2, and 3, that interact with each other. Together, they form an isolated system, so that the total energy $E_1 + E_2 + E_3$ is constant. Let's see what happens to system 1, when it undergoes two types of interactions: it receives heat energy from system 2 and does work on system 3. The situation is schematically described in Fig. I–11. The conservation of energy can be written as

$$E_{1i} + E_{2i} + E_{3i} = E_{1f} + E_{2f} + E_{3f}.$$

Rearranging terms and using the Δ notation, we have

$$\Delta E_1 = -\Delta E_2 - \Delta E_3. \tag{I–18a}$$

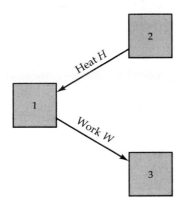

Figure I–11 Three systems. System 1 receives heat from system 2 and does work on system 3.

Let the heat energy that system 2 gives to system 1 be denoted by H, that is, $H = -\Delta E_2$ (the minus sign because the energy of system 2 decreases upon giving heat to system 1). Let the work done on system 3 by system 1 be denoted by W, that is, $W = \Delta E_3$ (a positive sign because the energy of system 3 increases after work is done on it). Finally, denote the change in energy of system 1 by ΔE, that is, $\Delta E = \Delta E_1$. Substituting these abbreviations into Eq. (I–18a), we obtain the first law of thermodynamics:

$$\Delta E = H - W. \qquad \text{(I–18b)}$$

In words, Eq. (I–18b) says that the increase in energy of a system (system 1) equals the heat energy received by that system minus the work expended by that system. If the system gives heat instead of receiving it, Eq. (I–18b) can still be used, with H negative. Likewise, if the system is worked on by another system instead of vice versa, W is negative.

Problem I–9 Pumping Iron with Heat

A container holds 0.015 kilograms of oxygen gas at 35 degrees C. A movable piston with a 5-kilogram block sits on top of the gas, 0.3 meters high, as shown in Fig. I–12a. Now, 15 calories of heat is added to the oxygen gas. The gas expands and lifts the block to a height of 0.6 meters, as shown in Fig. I–12b. What is the new temperature of the gas?

Solution: Apply the first law of thermodynamics, Eq. (I–18b). The heat added is $H = 15$ calories $= 15$ calories $\times$ 4.2 joules/calorie $= 63$ joules. The work done by the gas in lifting the block is the increase in gravitational energy of the block,

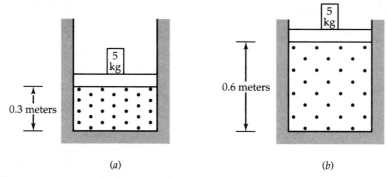

(a)   (b)

Figure I–12 (*a*) Gas inside a container, with a piston and mass on top. (*b*) The gas
expands and lifts the mass.

$$W = mg \, \Delta h = 5 \text{ kg} \times 9.8 \text{ m/s/s} \times (0.6 - 0.3)\text{m} = 14.7 \text{ joules.}$$

Thus the increase in energy of the gas is $\Delta E = H - W = 63$ joules $- 14.7$
joules $= 48.3$ joules. This energy goes into increasing the heat energy and
the gravitational energy of the gas. However, the latter is tiny, because the
average gas molecule rises only about 0.15 meters, leading to a total
increase in gravitational energy of the gas of only 0.015 kg $\times$ 9.8 m/s/s $\times$
0.15 m $= 0.02$ joules. Therefore, we can neglect the change in gravitational energy of the gas and assume that all 48.3 joules goes into increasing the
heat energy of the gas. To find the resulting increase in gas temperature,
apply Eq. (I–15b), with $T_i = 35C$, $\Delta E = 48.3$ joules, $M = 0.015$ kilograms,
and $c = 655$ joules per kilogram per degree C:

$$(T_f - 35 \text{ C}) = \frac{48.3 \text{ joules}}{0.015 \text{ kg} \times 655 \text{ j/kg/C}}$$

or

$$T_f = 35C + 4.9C = 39.9C.$$

G. THE CONSERVATION OF ENERGY AND THE LIMITED LIFETIME OF THE WORLD

Although discovery of the law of the conservation of energy brought order to
the world, it helped dash belief in a world without end. The principal source
of energy for life on earth is the sun. According to the conservation of energy,
the sun cannot create energy from nothing. Thus the sun cannot continue radi-
ating heat and light into space forever. To do so would require an infinite sup-

ply of energy, and the sun, and all bodies, contain only a limited amount of energy.

In 1862, the Scottish physicist William Thomson (1824–1907), introduced earlier as Lord Kelvin, estimated that the sun could not have illuminated the earth for more than about 100 million years. Scientists had proposed that the sun might replenish its energy loss by chemical energy, or by the constant influx of meteors, bringing in outside kinetic energy on a continuing basis, or by a slow shrinkage, liberating its own gravitational energy. Thomson debated all of these theories. The chemical energy of the sun would have been depleted in only 1000 years; the amount of matter needed in the meteoric-influx hypothesis would have deflected planetary orbits to a noticeable degree; and in the gravitational-shrinkage hypothesis, the sun could have produced energy at its current rate for only the last 10 million years. Thomson argued that the sun was running on only the heat energy contained in its gases. The sun was unavoidably cooling and dimming and could not have radiated for more than 100 million years into the past. This was the age of the sun. By inference, it was also the age of the earth as we know it.

Thomson's calculations flew in the face of the Uniformitarian school of geology. This school, founded by the great nineteenth-century geologist Charles Lyell (1797–1875), argued that the forces in nature were constant and uniform, that the earth in the past was pretty much as it is today. Lyell and others claimed that the age of the earth could not be given any absolute bounds and refused to consider whether the earth even had an origin. The appreciation of the limited lifetime of the sun was a crucial step in humankind's gradual realization of the mortality of the universe, of the limited supplies of available energy, and of the inevitable running down of the cosmos. We will have much more to say about this in the next chapter.

As a footnote, we might mention that while Thomson's recognition that the sun had a limited lifetime was extremely important, his actual estimate of that lifetime was too low. Thomson didn't know about nuclear energy, discovered in the twentieth century. The available nuclear energy in the sun is several hundred times larger than its heat energy, allowing the sun to have shined for something like 10 billion years instead of 100 million. Nevertheless, even the sun's nuclear energy is limited. The sun has not always been shining, and it will one day cease to shine.

H. REACTIONS TO POSSIBLE VIOLATIONS OF THE CONSERVATION OF ENERGY

The law of the conservation of energy is one of the most sacred principles of physics. At times, though, it has seemed to be violated.

One such occasion happened in the late 1920s and involved the emission of electrons from radioactive atomic nuclei. Since the early twentieth century, it had been known that the centers of atoms, or atomic nuclei, can disintegrate, spitting out pieces of themselves. In one type of disintegration, called beta decay, the nucleus emits an electron, which is a relatively lightweight and negatively charged subatomic particle. In the 1920s experiments in nuclear physics showed that the energy of an atomic nucleus can change only in very definite amounts, rather than in a continuous way. For example, a uranium nucleus might be able to change its energy by 1.6×10^{-13} joules or by 2.8×10^{-13} joules, but not by any intermediate values such as 1.7×10^{-13} joules or 1.8×10^{-13} joules. This interesting observed behavior was well described by a new theory in physics called quantum mechanics, which we will discuss more in Chapter IV.

Now we come to the conservation of energy. When an atomic nucleus emits a particle, the law of conservation of energy demands that the energy carried away by the escaping particle equals the energy lost by the nucleus. Since an atomic nucleus can change its energy only in particular, discrete amounts, scientists expected that the escaping electron in a beta decay would also necessarily have one of these particular amounts of energy. Measurements in 1927 showed otherwise. The electrons emitted in beta decay came with a *continuous* range of energies. It was as if you had just tuned a piano and expected to hear the notes C, C#, D, D#, E, and so on, but instead heard all those notes plus thousands of notes in between.

Physicists were profoundly puzzled and disturbed by these results. The great Danish atomic physicist Niels Bohr (1885–1962), winner of the Nobel prize in 1922, bravely proposed that the law of the conservation of energy was being violated on the atomic scale. In his Faraday lecture of May 8, 1930, Bohr said:

> At the present stage of atomic theory we have no arguments, either empirical or theoretical, for upholding the energy principle in the case of beta ray disintegrations . . . Just as the account of those aspects of atomic constitution essential for the explanation of the . . . ordinary properties of matter implied a renunciation of the classical idea of causality, the features of atomic stability . . . may force us to renounce the very idea of energy balance. [*Journal of the Chemical Society*, vol. 135, p. 349 (1932)]

Bohr was not happy with abandoning the law of conservation of energy, even in the microscopic domain of the atom. In an unpublished article in 1929 titled "Beta Ray Spectrum and Energy Conservation," he wrote that "the loss of the unerring guidance which the conservation principles have hitherto offered in the development of the atomic theory would of course be a very disquieting prospect." (Niels Bohr library, American Institute of Physics.)

Leading physicists split over whether to renounce the law of the conservation of energy (and all the certainty implied by that law). One of the founders of quantum theory, Werner Heisenberg (Nobel prize 1932), sided with Bohr. On the other side, Wolfgang Pauli (Nobel prize 1945), Ernest

Rutherford (Nobel prize 1908), and Paul Dirac (Nobel prize 1933) were reluctant to part with energy conservation even though it appeared to be violated. In a letter to Bohr on November 19, 1929, Rutherford wrote

> I have heard rumors that you are on the war path and wanting to upset Conservation of Energy, both microscopically and macroscopically. I will wait and see before expressing an opinion, but I always feel "there are more things in heaven and Earth than are dreamed of in our Philosophy." [*Niels Bohr, A Centenary Volume*, ed. A. P. French and P. J. Kennedy (Harvard University Press: Cambridge, Massachusetts, 1985), p. 200]

Pauli wrote to the physicist Oskar Klein in a letter dated February 1929, "With his considerations about a violation of the energy law, Bohr is on a completely wrong track." [*Inward Bound*, by Abraham Pais (Oxford University Press: Oxford, 1988), p. 309]

Pauli, in fact, proposed a solution to the troubling experimental results of beta decay, a solution preserving the law of the conservation of energy. In late December 1930, he suggested that there must be another particle emitted in addition to the electron. The *sum* of the energies of the *two* escaping particles would come at only particular values, as required by the conservation of energy and the quantum nature of the atomic nucleus, but the *individual* energies of the electron and the new particle could vary continuously, in accordance with the observations. The hypothesized particle would have to be practically undetectable, since it had never been seen. It would have to be nearly massless, electrically neutral, and almost incapable of producing any effect on ordinary matter. Twenty-five years later, Pauli's predicted particle, named the neutrino, was discovered and his faith in the conservation of energy upheld.

Discussion Questions I–13

Do you think the law of the conservation of energy is just a matter of the definition of energy and invented rules, or does it mean more? Does the law say something about nature or about the human mind? Do you think scientists will always be able to find new forms of energy and new particles to make the law of conservation of energy continue to hold true? Why is this conservation law so important to scientists?

Readings

Lucretius, *De Rerum Natura*, trans. W. H. D. Rouse and M. F. Smith, Loeb Classical Library (Harvard University Press: Cambridge, Massachusetts, 1982) Books I–III.

Benjamin Thompson (Count Rumford), "Convection of Heat," (1798) and "Inquiry Concerning the Source of Heat Which is Excited by Friction" (1798) in *A Source Book in Physics*, ed. William H. Magie (McGraw-Hill: New York, 1935).

James Joule, "On Matter, Living Force, and Heat," (1847) in *Kinetic Theory*, vol. I, ed. S. G. Brush (Pergamon Press: Oxford, 1965).

WILLIAM THOMSON
(BARON KELVIN OF LARGS)

William Thomson (1824–1907) was born in Belfast, Ireland. His mother died when he was 6 years old. In his early years, Thomson was educated at home by his father, James Thomson, a professor of engineering and mathematics. In 1834, William and his brother went to Glasgow University, and in 1841 Thomson entered Cambridge University. While at Cambridge, Thomson was awarded silver sculls for winning the university championship in crew.

The French school of science, involving strong mathematical analysis, was influential in Glasgow, and Thomson went to Paris after his graduation from Cam-

bridge University. In 1845, Thomson returned to Scotland, where his father shrewdly campaigned to land his son a recently vacated professorship in natural philosophy (science) at the University of Glasgow. The younger Thomson received the professorship in 1846, at the age of 22. This post he held for the rest of his life.

In his work on heat and the laws of thermodynamics, Thomson was at first hampered by his conception of heat as a conserved substance, rather than a molecular motion that could be completely converted into another form of energy. Many other scientists of the time had the same misconception. After the study of both actual and theoretical heat engines, Thomson proposed in 1853 the second law of thermodynamics, in the following form: "It is impossible, by means of inanimate material agency, to derive mechanical effect from any portion of matter by cooling it below the temperature of the coldest of the surrounding objects." Thomson's proposition was an assertion—not a derivation based on probability theory.

In the 1850s, Thomson applied his understanding of heat and thermodynamics to discuss the cooling of the earth and sun. He argued that if those bodies had originated as hot molten balls, as believed, then their rate of cooling must have been much higher in the past, when temperature differences were higher. Consequently, geological activities such as winds and volcanoes, ultimately driven by heat flow, would have been more violent in the past. These considerations contradicted the common geological wisdom, which held that geological conditions had not substantially changed over time. Thomson, mild mannered by nature, was unwittingly thrown into a controversy.

In addition to his theoretical work on heat, electricity, and magnetism, Thomson was keenly interested in the design of new instruments for measurement. In fact, he was exceptional in his combined abilities at theoretical work and instrumentation. In the mid-1850s, Thomson was recruited to serve on a project to lay an underwater telegraph cable between Ireland and Newfoundland, the first of its kind. The project was backed by a group of British industrialists. Thomson invented a very sensitive device to measure tiny currents in the long wire, but his device was rejected by another scientist on the project, jealous of him. On the third cable laid, in 1865, Thomson's device was finally used, and the project was a great success. For this victory, Thomson was celebrated by the British financiers, became rich himself, and was knighted by Queen Victoria.

Thomson's guiding philosophy was that all physical phenomena should be reducable to mechanical models. He frequently championed other scientists' ideas. Thomson loved the sea, and he spent a lot of time on his yacht. From these interests came patents on a new kind of compass, a device for calculating tides, and depth-finding equipment.

Thomson had two marriages and no children. He was buried in Westminster Abbey, London.

CHAPTER 2

The Second Law of Thermodynamics

Since energy is conserved, why does energy have to be fed to a car to keep it from stopping? Since energy is conserved, why does hot soup cool and ice cream melt? Why does smoke fill up a room but never crowd into one corner? Why does time flow forward but never backward? These diverse phenomena all illustrate a profound idea of physics, the second law of thermodynamics: the level of disorder in the world is rising, relentlessly and irreversibly.

Historically, the notion of a stable and unchanging universe has always been appealing, and the second law upset many people, including scientists, when it was discovered in the mid-nineteenth century. The second law says that some processes in nature are one-way arrows, never going backward, never returning the world to its initial condition. The machines are running down. The universe, on average, is dissipating itself.

A. REVERSIBLE AND IRREVERSIBLE PHENOMENA

To understand the second law of thermodynamics, we will begin with a discussion of reversible and irreversible processes. Let's return to our pendulum and place it in a box without air. Assume further that the joints of the pendulum are perfectly smooth. Release the pendulum from a height, and it will keep swinging forever, converting its gravitational energy to kinetic energy and back again endlessly. This is a *reversible process*. A movie of the swinging pendulum would look the same whether the movie was played forward or backward. The pendulum always repeats its cycle exactly, returning to the exact height from which it was released.

Now, put some air in the box, a more normal situation. What happens, of course, is that the pendulum gradually slows down and stops, and the air gradually heats up as it absorbs the pendulum's energy. The energy of the pendulum has gone into heat. This is what happens to all pendulums and machines that are not periodically wound up or reset or reenergized in some way. There is always at least a bit of friction that slows them down and stops them.

Have we ever seen a motionless pendulum suddenly start swinging? No. Similarly, we've seen vases fall off tables and break into pieces, but we've never seen the fractured pieces of a vase suddenly pull themselves together, form a vase, and catapult upward onto a table. Eggs often break, but never re-form; skywriting fades and never comes back; unattended rooms gather dust, but do not get clean. These are examples of *irreversible processes*. They always go in one direction, but not in the opposite.

The law of conservation of energy does not forbid the pendulum to start swinging; the required energy could be supplied by the motions of the molecules in the surrounding air. Nor does the conservation of energy forbid the broken vase on the floor to re-form itself suddenly and leap to the table; the required energy could be supplied by the heat and vibrations in the floor. These reverse processes are all possible according to the law of conservation of energy, but they never seem to happen. Why? Evidently, there is at work some additional principle that makes some phenomena irreversible. Such a principle must be universal, since it applies to so many things.

That many phenomena in nature are irreversible is an expression of the *second law of thermodynamics*. As mentioned before in Chapter I, the conservation of energy is often called the first law of thermodynamics. If there were no second law, the universe would be like a giant clock that never runs down. (This simile was often used in the seventeenth and eighteenth centuries, after the mechanical laws of Galileo and Newton but before the discovery of the second law of thermodynamics.) If there were no second law, pendulums would keep swinging. Planets would repeat their trajectories without loss. A new star would form for every star that burned out. Although successive snapshots of the universe would show change, the changes would cycle, exactly repeating themselves, so that a *time-lapse* picture of the universe would look the same from one epoch to the next.The universe, on average, would not change.

The second law was first stated in 1852 by William Thomson (Lord Kelvin):

> 1. There is at present in the material world a universal tendency of the dissipation of mechanical energy. 2. Any restoration of mechanical energy, without more than an equivalent of dissipation [loss of mechanical energy elsewhere], is impossible in inanimate material processes, and is probably never effected by means of organized matter, either endowed with vegetable life or subjected to the will of an animated creature.
> 3. Within a finite period of time past, the earth must have been, and within a finite period of time to come, the earth must again be, unfit for the habitation of man as at present constituted, unless operations have been or are to be performed, which are impossible under the laws to

which the known operations going on at present in the material world, are subject. ["Mathematical and Physical Papers," of William Thomson (Cambridge University Press: Cambridge, England, 1882), vol. I, p. 514]

Discussion Question II–1

Why might a physicist, in stating a new law of physics, include references to "life," the "will of animated creatures," and the necessary conditions for the "habitation" of earth by man?

There are many equivalent statements of the second law of thermodynamics: isolated systems inevitably become less organized; the *usable* energy in an isolated system is constantly decreasing; a system naturally attempts to distribute its energy equally among all of its parts; mechanical energy, on average, degrades into heat; heat naturally flows from hot places to cold; isolated machines cannot remain in perpetual motion; entropy (disorder), on average, increases. These statements all express the second law of thermodynamics. Physicists believe that it is the second law of thermodynamics that defines the direction of time, that distinguishes the past from the future.

Why is there a second law of thermodynamics? To answer this question, we will delve into the theory of probability. We will show that the second law is not really a law but a statement of probabilities. The odds strongly favor the slowing and stopping of a pendulum. The odds say the machines must run down.

B. STATES OF A SYSTEM AND PROBABILITY OF CONFIGURATIONS

1. An Introduction to Probability Theory

Much of the theory of probability can be illustrated with a simple example. If we have a pair of dice, what is the probability of rolling a 7? And what do we mean by probability?

We will answer the second question first. The probability of getting a number is the *fraction* of times you get that number in a large number of rolls. For dice, and for many other situations, we can compute the probabilities in advance, without doing a large number of trials. To figure out the probability of rolling a certain number, note that there are 36 possible combinations of the two dice. (There are 6 numbers that can come up on the first die, and *each* of these can be paired with 6 numbers on the second die, leading to $6 \times 6 = 36$ possible combinations.) Since all 6 faces of each die are identical in shape and equally likely to come up, each of the 36 possible combinations is equally likely to come up.

Let's look at each of the 36 combinations and see what sum it produces, shown in Table II–1a. For example, there are 3 combinations that produce a sum of 4: die A = 1 and die B = 3, die A = 2 and die B = 2, and die A = 3 and die B = 1. We will denote these combinations as (1, 3), (2, 2), and (3, 1), respectively.

We can now count how many combinations (rolls) make each sum (Table II–1b). The probability of getting any particular sum is the *fraction* of rolls that give that sum, that is, the number of combinations that give the sum divided by the total number of possible combinations, 36. The probability of each sum

TABLE II–1A: COMBINATIONS OF TWO DICE

Die A	Die B	Sum
1	1	2
1	2	3
1	3	4
1	4	5
1	5	6
1	6	7
2	1	3
2	2	4
2	3	5
2	4	6
2	5	7
2	6	8
3	1	4
3	2	5
3	3	6
3	4	7
3	5	8
3	6	9
4	1	5
4	2	6
4	3	7
4	4	8
4	5	9
4	6	10
5	1	6
5	2	7
5	3	8
5	4	9
5	5	10
5	6	11
6	1	7
6	2	8
6	3	9
6	4	10
6	5	11
6	6	12

TABLE II–1B: Probabilities of Sums

Sum	Number of Combinations Making Sum	Probability of Sum
2	1	1/36
3	2	2/36
4	3	3/36
5	4	4/36
6	5	5/36
7	6	6/36
8	5	5/36
9	4	4/36
10	3	3/36
11	2	2/36
12	1	1/35

in the first column of Table II–1b is shown in the last column of that table. As you can see, getting a 2 is relatively improbable because only one combination of the two dice produces this sum, (1, 1). Thus, the probability of rolling a 2 is only 1/36. Rolling a 4 is more probable because there are three combinations leading to a 4, as discussed. The probability of a 4 is therefore 3/36 = 1/12. Getting a 7 is even more probable because there are 6 combinations that will yield 7: (1, 6), (2, 5), (3, 4), (4, 3), (5, 2), and (6, 1), leading to a probability of rolling a 7 of 6/36 = 1/6.

If the dice were thrown millions of times, the sum of 2 would come up 1/36 of the time, the sum of 4 would come up 1/12 of the time, and the sum 7 would come up 1/6 of the time. It is important to realize, however, that on a *small* number of throws, there could be large departures from these results. Probabilities refer only to averages and become less and less meaningful as the number of cases becomes smaller and smaller, just as the average length of a pregnancy accurately applies to a large population, but does not allow an individual woman to plan a long journey near her due date.

You can test the above analysis. Roll a pair of dice 20 or 30 times and record the sum of each roll. Then make a table of how many times each sum comes up. Have a dozen friends do the same and combine all your results. For good averages, you should have at least a few hundred rolls altogether. Now compute the fraction of times each sum comes up and compare to Table II–1b. There should be a fairly close, but not exact, agreement. With a much larger number of rolls, the agreement would improve.

Problem II–1: Boys and Girls

If a family has three children, what is the probability that they are all the same gender?

Solution: Let B stand for boy and G stand for girl. Since there are two possibilities for each child, the total number of different combinations for three children is 2 × 2 × 2 = 8. The eight possible combinations are BBG, BGB, BBB, BGG, GBG, GGB, GGG, and GBB. Two of the eight combinations have all boys or all girls. Thus, the probability of getting one of these outcomes is 2/8 = 1/4. In other words, one fourth of all three-children families should have all boys or all girls.

Some new terminology will help us apply these ideas to more general situations. Each distinct combination of the dice is called a "state" of the system, and each possible sum is called a "configuration" of the system. For the two dice, there are 36 possible states and 11 possible configurations. To restate in these terms what we have learned from the dice, the probability of a certain configuration of a system is the fraction of states that lead to that configuration, assuming that all states are equally likely.

2. A Pendulum in Air

Let's again consider a pendulum swinging in a box of air and see what the odds say about where the energy of the system should go—into the pendulum or into the air. We will first consider a very simple example so that we can figure out everything without too much trouble.

Suppose that our simplified system consists of a box containing a pendulum and three identical molecules of gas (Fig. II–1). Assume the joints of the pendulum and the walls of the box are frictionless. Then the total energy of

Figure II–1: Pendulum inside a box containing three molecules of gas.

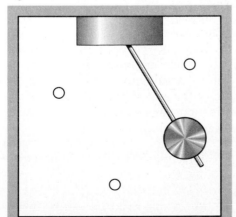

the system must remain in the pendulum and in the molecules. That total energy will always be divided in some way between the pendulum and the gas. What is the probability of each division?

Let us assume that the system has a total of 11 units of energy, $E_{tot} = 11$ units. It is unimportant whether the unit is a calorie or a joule or whatever. Denote the energy of the gas by E_g and the energy of the pendulum by E_p. The energy in the gas E_g is the *sum* of the energies of the three gas molecules. Conservation of energy requires the condition that $E_g + E_p = E_{tot} = 11$.

There are many ways that the total energy of the system E_{tot} could be divided between the gas and the pendulum. For example, the division could be $E_g = 3$ and $E_p = 8$, or it could be $E_g = 7$ and $E_p = 4$, and so on. We will assume for simplicity that each part of the system can have only whole units of energy. Thus, one of the molecules could have 3 units of energy, but not 4.7.

Now, what we want to know is the probability of each possible division of the energy between the pendulum and the gas, or, in other words, the probability of each value of E_g. This is analogous to asking for the probability of each sum of the dice. Here, what is summed are the energies of the three molecules to produce each value of E_g. For the dice, there were 11 different possible sums; for the gas, there are 12, since E_g can have any value from 0 to 11, the total energy in the system. To figure probabilities, we have to count the number of states of the system producing each possible value of E_g.

How is a state of the system specified? For the dice, it was specified by giving the number on each of the two dice. For our system of the pendulum and gas, a state is specified by giving the amount of energy in each of the three gas molecules. The energy in the pendulum is then also known, since the total energy of the system is fixed, $E_p = E_{tot} - E_g = 11 - E_g$. (For a realistic gas, additional factors might be involved in specifying a state, such as the positions of the molecules, but such complications do not qualitatively change our analysis.)

Each value of E_g corresponds to a configuration of the system. For a given value of E_g, that is, a given total energy in the gas, there can be different ways of distributing this energy among the three molecules. For example, a value of $E_g = 2$ could be achieved by putting 1 unit of energy in two molecules and none in the third, or 2 units of energy in one molecule and none in the other two molecules. Each different way of distributing this energy among the molecules corresponds to a different state of the system, just as each sum of the dice can be produced by several combinations, or states, of the two dice.

Before counting states, we must recognize a difference between the dice and the molecules. We could distinguish the two dice, labeling one "A" and one "B." A die is big, and a tiny nametag can be attached to it without hardly affecting its appearance, leaving it identical to other dice for all practical purposes. On the other hand, identical molecules are indistinguishable. They cannot be labeled without changing their identity. Thus, two dice can be distinguished from each other, while two oxygen molecules cannot. The result is that while a state of the dice with die A reading 3 and die B reading 5 is distinct from a state with A reading 5 and B reading 3, a state of the gas with molecule A having 2 units of energy and molecule B having 3 units cannot be distinguished from a state with A having 3 and B having 2. There just aren't any labels on the molecules, allowing one to be called "A," another "B," and so on. All that can be specified is how many molecules have each amount of energy.

Now, let's count states. We can use the same notation we used for the dice. For example, (3, 2, 0) denotes a state in which the first gas molecule has 3 units of energy, the second has 2 units of energy, and the third has no energy, for a total energy in the gas $E_g = 3 + 2 + 0 = 5$. Since the molecules are indistinguishable, the state (3, 2, 0) is not different from the state (2, 0, 3) or any other arrangement of 3, 2, and 0. This state *is* different, however, from the state (2, 2, 1), in which two molecules have 2 units of energy and one molecule has 1 unit. Note that both (3, 2, 0) and (2, 2, 1) have the same gas energy, $E_g = 5$, and so correspond to the same configuration of the system. Table II–2 lists the different states of the system for each value of total energy in the gas, E_g, ranging from 0 to 11 units. As you can see, the number of states increases as the energy

TABLE II–2: States Available to a Three-Molecule Gas for Different Energies

Energy of Gas, E_g	Possible States	Number of States
0	(0, 0, 0)	1
1	(1, 0, 0)	1
2	(2, 0, 0) (1, 1, 0)	2
3	(3, 0, 0) (2, 1, 0) (1, 1, 1)	3
4	(4, 0, 0) (3, 1, 0) (2, 2, 0) (2, 1, 1)	4
5	(5, 0, 0) (4, 1, 0) (3, 2, 0) (3, 1, 1) (2, 2, 1)	5
6	(6, 0, 0) (5, 1, 0) (4, 2, 0) (4, 1, 1) (3, 3, 0) (3, 2, 1) (2, 2, 2)	7
7	(7, 0, 0) (6, 1, 0) (5, 2, 0) (5, 1, 1) (4, 3, 0) (4, 2, 1) (3, 3, 1) (3, 2, 2)	8
8	(8, 0, 0) (7, 1, 0) (6, 2, 0) (6, 1, 1) (5, 3, 0) (5, 2, 1) (4, 4, 0) (4, 3, 1) (4, 2, 2) (3, 3, 2)	10
9	(9, 0, 0) (8, 1, 0) (7, 2, 0) (7, 1, 1) (6, 3, 0) (6, 2, 1) (5, 4, 0) (5, 3, 1) (5, 2, 2) (4, 4, 1) (4, 3, 2) (3, 3, 3)	12
10	(10, 0, 0) (9, 1, 0) (8, 2, 0) (8, 1, 1) (7, 3, 0) (7, 2, 1) (6, 4, 0) (6, 3, 1) (6, 2, 2) (5, 5, 0) (5, 4, 1) (5, 3, 2) (4, 4, 2) (4, 3, 3)	14
11	(11, 0, 0) (10, 1, 0) (9, 2, 0) (9, 1, 1) (8, 3, 0) (8, 2, 1) (7, 4, 0) (7, 3, 1) (7, 2, 2) (6, 5, 0) (6, 4, 1) (6, 3, 2) (5, 5, 1) (5, 4, 2) (5, 3, 3) (4, 4, 3)	16

in the gas increases, reflecting the simple fact that more energy allows more different arrangements of that energy among the three molecules.

Considering now the total system of gas plus pendulum, we can list the number of states available to the system for each of the 12 possible divisions of the total energy (Table II–3). There are 83 possible states of the system, analogous to the 36 possible states of the two dice.

What is the probability of each division of energy (configuration)? For the dice, any of the 36 possible combinations was equally likely with every roll. Every roll of the dice allowed the system to go from an initial state to any other state with equal likelihood, regardless of the initial state. For the gas and pendulum, the situation is a little more complicated. Random collisions of the molecules with each other and with the pendulum correspond to rolls of the dice. These collisions transfer energy between different parts of the system and allow it to go from one state to another. However, these collisions transfer only a little energy at a time. If we do not wait for many collisions between the molecules and the pendulum, then the system will not have had time to evolve to any possible state from its initial state. Only those states with a distribution of energy close to that of the initial state are possible. In such a situation, the probability of each state of the system is not equal. States that require only a small number of molecular collisions to be reached from the initial state are highly probable, while states that require a large number of collisions are highly improbable. The probability of moving to a new state then depends on the *initial* state of the system, unlike the case of the dice.

On the other hand, if we wait long enough for many molecular collisions to occur, then the full energy of the system can have been shuffled and reshuf-

TABLE II–3: States Available to a System of a Three-Molecule Gas and Pendulum with 11 Units of Total Energy, for Various Divisions of the Energy

Energy of Gas, E_g	Energy of Pendulum, E_p	Number of States of System (from Table II–2)	Probability
0	11	1	1/83
1	10	1	1/83
2	9	2	2/83
3	8	3	3/83
4	7	4	4/83
5	6	5	5/83
6	5	7	7/83
7	4	8	8/83
8	3	10	10/83
9	2	12	12/83
10	1	14	14/83
11	0	16	16/83

fled among all of its parts. The system can move to any new state from any initial state with equal probability. The situation is similar to the way money is redistributed in a small community. If a person with a huge bankroll moves to town, for the first few days it is more likely that a dollar in someone's pocket came directly from the wealthy newcomer than from anybody else. After a few months, however, the new money has been well circulated, and a dollar in someone's pocket could just have likely come from the butcher or the mail carrier or a neighbor.

Once sufficient time has passed and all states are equally likely, the pendulum and gas behave similarly to the dice: All possible states are equally likely, and the probability of a particular configuration is just the fraction of states that produce that configuration. These probabilities are given in the last column of Table II–3.

Problem II–2: Collision Time Scales

Suppose a pendulum is set swinging in a gas and collides with 10^{20} molecules of gas per second. On some of these collisions, the pendulum will gain energy, on some it will lose. Suppose that, on average, the pendulum loses a fraction 10^{-23} of its initial energy to a molecule on each collision. Estimate how long it takes the initial energy of the pendulum to be completely transferred to the gas molecules.

Solution: After $1/10^{-23} = 10^{23}$ collisions, all of the energy of the pendulum can have been transferred to the gas. This requires

$$\frac{10^{23} \text{ collisions}}{10^{20} \text{ collisions/second}} = 1000 \text{ seconds.}$$

After this period of time, the initial energy of the system has been completely reshuffled, and the system is equally likely to be found in any of its possible states.

After only 1 second, on the other hand, only 0.1% of the initial energy of the pendulum is likely to have been transferred to the molecules, and states of the system with more than 0.1% of the initial energy in the molecules are very unlikely.

To gain a better understanding of the meaning of Table II–3, let's start our simple system in some initial state (for example, with all the energy in the pendulum and none in the gas) and wait long enough for many molecular collisions to occur. Then we look again at the system. Each possible state is equally likely, so the probability of each configuration is the fraction of states that produce that configuration. These probabilities are shown in the last column of Table II–3. As you can see, the probabilities increase in the direction of taking energy out of the pendulum and putting it into the gas. In other words, after a sufficient period of time, the odds favor finding the pendulum at rest and all the energy in the gas, regardless of the initial state.

To be quantitative, Table II–3 shows that finding all the energy in the gas is 16 times more likely than finding it all in the pendulum. Finding all the energy in the pendulum at some later time is not impossible; it is just improbable compared to other possibilities. Intermediate situations, with some energy in the gas and some in the pendulum, are also possible, of course, with the probabilities shown in the table.

The second law of thermodynamics says that a swinging pendulum will slow down and come to rest. As illustrated in Table II–3, the second law is a statement of probabilities, not certainties. Although the odds are against it, a pendulum can sometimes gain energy from the surrounding air molecules.

3. Evolution to More Probable Configurations

The real world evolves. How do probabilities relate to evolution? What we have done so far is to evaluate the probability for energy to be divided in various ways between the gas molecules and the pendulum. The odds favor finding the system with most of its energy in the gas. Suppose we start the system with most of its energy in the pendulum. Then the pendulum will slow down, gradually transferring its energy to the gas and moving to a configuration with the highest probability. But how did the system discover that it was started in an unlikely state? How and why does a system *evolve* in the direction of more probable configurations?

To answer these questions, we will do an experiment. Consider the system shown in Fig. II–2a. Here we have a ball bouncing around in a series of chambers. The ball can get from one chamber to the next if it passes through a hole in the wall separating the two chambers. The holes are all the same size and arranged so that each wall has twice as many holes as the wall to its left. Thus, whatever chamber the ball is currently in, it is twice as likely to next enter the chamber to the right as to the left. More generally, each chamber corresponds to a configuration of the system, and the odds increase in the direction of chambers to the right, analogously to their increase in the direction of more energy in the gas in Table II–3.

To begin our experiment, we toss the ball into chamber 2, with a random direction, and follow its path. The dashed lines show a typical path of the ball as it bounces off the walls, occasionally encountering a hole and entering a new chamber. How does this system evolve?

We could build such a system and do the experiment. Instead, we will simulate it. The only equipment we need for our simulated experiment is a

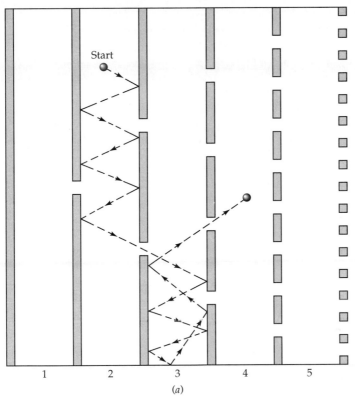

Start

1   2   3   4   5

(a)

Figure II–2: (a) Ball bouncing around inside a box with a series of chambers. The right
wall of chamber 1 has 1 hole in it, through which the ball can pass. The
right wall of chamber 2 has two holes in it; the right wall of chamber 3 has
four holes in it; and so on. The ball is thrown into chamber 2 with a
random direction. Dashed lines indicate the subsequent motion of the ball.
The box lies horizontally on a table.

pencil, a piece of paper, and a die. We draw a series of horizontal lines with
equally spaced dots on each line, as shown in Fig. II–2b. Each dot represents a
chamber, and each higher horizontal line shows a snapshot of the chambers at
a later time, when the ball has moved to a new chamber. As the ball moves
from one chamber to the next, it moves to the corresponding dot on the next
higher horizontal line.

Start the ball off in some initial chamber in the bottom horizontal line. (In
the figure, we have circled the initial chamber. Any dot can be chosen as the
initial chamber, as long as there is room on the page or blackboard to follow
the simulated experiment.) To find the next location of the ball, roll a die. If 1
or 2 comes up, move the ball to the dot to the left, on the next horizontal line
up. If 3, 4, 5, or 6 comes up, move the ball to the right. The probabilities here

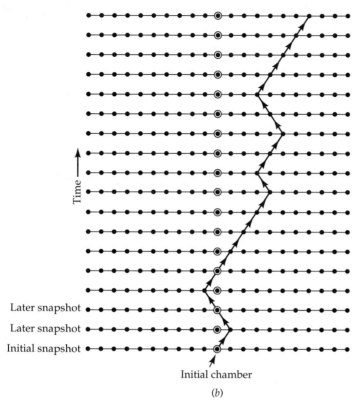

Later snapshot

Later snapshot

Initial snapshot

Initial chamber

(b)

Figure II–2: (b) A representation of the path of the ball, started off in some initial chamber. Each horizontal line shows the position of the ball at a certain time. Each dot on one of the horizontal lines stands for one of the chambers.

correspond exactly to those in the real experiment with holes in the walls between chambers. Since there are twice as many numbers that move the ball to the right as to the left, the probability of moving to the right is twice as great. The randomness of the toss of the die mimics the randomness of the path of the ball. That path will vary from one experiment to the next, depending on how the ball is initially tossed into the first chamber.

Roll the die 15 times or more, connect the successive locations of the ball, and you will see how your system evolves in time.

Figure II–2b shows the results of a typical simulated experiment. The precise path of the ball from one chamber to the next will most likely be different in your own experiment and in most new experiments, because there is randomness in the toss of the die, but the general features of all experiments should be the same. You should find, as in Fig. II–2b, an overall trend of motion to the right, even though there will be occasional excursions to the left.

In the experiment shown in Fig. II–2b, for example, the ball improbably finds itself to the left of where it started off after three rolls of the die. After 17 rolls, however, it is located 7 chambers to the right of where it began.

Our simulated experiment is exactly analogous to the real experiment. By bouncing back and forth off the walls of each chamber, the ball, on average, finds more holes to the right than to the left and thus, on average, advances to a new chamber in that direction. In this way, the greater probability of the configuration to the right results in an evolution of the system in that direction. Just as in the simulated system, the real system evolves toward configurations of increasing probabilities only *on average*. At times, the system will evolve toward a less probable configuration.

Also, as in the simulated system, *randomness is critical*. The ball was casually tossed into the first chamber, so that its initial direction was "typical." If, instead, the holes and chambers were analyzed ahead of time, it would be possible to carefully choose the initial direction of the ball so that it never encountered a hole, or even preferentially moved to the left. In this case, the ball would not have an equal chance of passing through any hole because its initial position and direction would have been set up in a special way. An element of randomness is necessary for a system to evolve to a more probable configuration, and indeed for the concept of probability to have meaning. This simply means that the initial conditions for the system must be average or typical, rather than special.

Returning now to the pendulum and gas, we can understand how the pendulum gradually transfers its energy to the gas. The trend of evolution follows the direction of increasing probabilities, although energy will sometimes be transferred in the other direction, from the gas to the pendulum. Collisions between the molecules and the pendulum correspond to the collisions between the ball and the chamber walls. Collisions constantly reshuffle the energy and allow the system to find all the states available to it, just as the back-and-forth collisions of the ball allow it to find all the available holes.

We can also see why it is essential for the initial positions and motions of the molecules to be random. We could always arrange the molecules so that they *never* collided with the pendulum. For example, we could place them all at the bottom of the box, headed precisely at right angles to one of the sides. Then they would keep bouncing back and forth from left to right but remain below the pendulum. For a special initial arrangement of the molecules, the system might not evolve at all. So, again, the system must have some randomness in order to evolve toward more probable configurations.

A pendulum swinging in a three-molecule gas, although a highly simplified model of reality, fully illustrates the second law of thermodynamics. If you understand that simple system and why it evolves as it does, then you understand the essence of the second law of thermodynamics. Everything else in this chapter will be an application of what you have just learned. First, however, we should see how our three-molecule gas relates to a more realistic gas.

4. Behavior of Realistic Systems with Large Numbers of Molecules

If we look again at Table II–3, we can see that although the probabilities do indeed increase one way in the direction of placing more energy in the gas and less in the pendulum, in accordance with the second law of thermodynamics, the increase is not large. Although the trend of evolution is toward the configuration with all the energy in the gas, there will be fairly large departures from this configuration from time to time, and the pendulum will behave somewhat erratically. For example, the probability is nearly 50% that at any moment the pendulum will have at least 3/11 of the total energy or more, since $10/83 + 8/83 + 7/83 + 5/83 + 4/83 + 3/83 + 2/83 + 1/83 + 1/83 = 41/83 = 49\%$. In other words, after sufficient time for the energy to be shuffled around, the pendulum will spend about half its time with 3/11 or more of the total energy. If we started our simple system with all the energy in the pendulum, we would see the pendulum slow down, then speed up, then slow down, and so on, in a jerky fashion—slowing down on average but in a very uneven way. Then, after it had nearly come to rest, it would start moving again (after a gas molecule collided with it), gaining some energy, losing some energy, and so on, spending about half its time with 3/11 or more of the total energy. Even if we *start* the system with the pendulum at rest and all the energy in the gas, the most probable configuration, the pendulum won't remain at rest. It will get kicked by a molecule, gaining energy, and continue to gain and lose energy in a random fashion, spending about half its time with 3/11 or more of the total energy. That is what the probabilities of Table II–3 mean.

Three-elevenths is a good fraction of the total energy. It would be clearly noticed in any realistic system. A cubic foot of air at room temperature, for example, has about 10^4 joules of energy in the random motions (heat) of its molecules. Three-elevenths of that amount of energy would kick a 1-pound pendulum half a mile high.

What is the meaning of the second law of thermodynamics if it can be violated so brazenly? And why have we never witnessed such a violation? The resolution of this paradox lies in the number of molecules of the gas. Our calculations were based on a gas of only three molecules, so that we could easily work out all the numbers. A realistic box of gas, however, has far more molecules. With many more molecules, the probabilities *overwhelmingly* favor putting all the energy in the gas. The detailed proof of this result is given in Appendix B. Roughly speaking, with a huge number of molecules, there is a huge increase in the number of possible states of the gas for even a little increase of energy of the gas, thus causing the probabilities to dramatically increase in the direction of putting more energy into the gas. The result of these considerations is that for a realistic situation with many molecules, a pendulum initially with all the energy would smoothly slow down, come to rest, and remain nearly at rest.

Appendix B, which extends our results to systems with a large number of molecules, demonstrates another important feature of the second law of thermodynamics: *An isolated system evolves toward a configuration in which each part of the system has the same amount of energy.* Thus, in a system with N parts and a total energy E_{tot}, each part will end up with an energy of about E_{tot}/N. In the case of the pendulum in a gas, this form of the second law says that the system will evolve until the pendulum has about the same amount of energy as a single molecule of gas—a very tiny fraction of the total energy for a system of many molecules.

In conclusion, the trend of evolution toward placing all the energy in the gas and none in the pendulum, worked out in detail for the three-molecule gas, is overwhelmingly favored for a realistic gas with a large number of molecules. This one-way trend toward more probable configurations is the essence of irreversibility and underlies the second law of thermodynamics.

5. The Direction of Time

In our experience with the world, nothing is more basic than our sense of time. We have a strong feeling for the forward direction of time, for the distinction between past, present, and future. Yet why do we have this feeling? In fact, the direction of time is connected to the second law of thermodynamics.

Discussion Questions II–2

Suppose you have a reel of motion picture film, showing a pendulum swinging back and forth, slowing down, and gradually coming to rest. You have a second copy of the reel, identical to the first but wound in the opposite direction, showing a pendulum first sitting at rest, then gradually swinging faster and faster. You invite some friends to look at the two films and ask them which film is being shown "forward" and which "backward." Will your friends agree with each other?

Discuss how your sense of the direction of time might be determined by the second law of thermodynamics. How do you know whether something is going in the forward or backward direction? Suppose a pendulum is at rest, in its most probable state. Would a motion picture of this pendulum define a direction of time?

Discuss how your brain might be aware of its own departure from the most probable state and thus define its own direction of time.

Suppose you lived in a universe where all systems were in their most probable state. Would there be a "direction of time?" How would the world seem to you?

Physicists believe that the direction of time is so clear because our universe *began* in a highly *improbable* configuration and has been evolving to more probable configurations, in a one-way fashion, ever since. Why it is that our universe began in such an improbable condition is an outstanding mystery.

C. MECHANICAL ENERGY AND HEAT

It is useful to contrast heat energy to kinetic and gravitational energy, both of which are called *mechanical energy*. The energy of a swinging pendulum is an example of mechanical energy. The kinetic energy of any large object, all of whose parts are moving together in bulk, is called mechanical energy. In the case of the swinging pendulum, the pendulum bob contains many molecules moving together in bulk. This energy is to be distinguished from *heat* energy, the energy of the random motions of molecules. The gravitational energy of an object is also called mechanical energy, since such energy can be converted into the bulk motion of the object (after the release of a pendulum, for example). The slowing down of a pendulum illustrates mechanical energy turning into heat energy.

All irreversible phenomena are equivalent to converting mechanical energy to heat. The following example shows why. One half of a box is filled with a gas and the other half is empty. A partition separates the two halves (Fig. II–3a). Now the partition is removed, and the gas soon fills the entire box (Fig. II–3b). This is an irreversible process because the gas molecules will never again congregate in the lower half of the box, unless some outside agent forces them to do so. Gas molecules tend to fill all the space available rather than congregate in a small portion of that space, as is apparent when smoke is released in a room.

To see that the irreversible process just described is equivalent to converting mechanical energy to heat, consider what would be necessary to

Figure II–3: (*a*) Molecules of gas confined to the bottom half of a box. (*b*) The separating partition has been removed, and the molecules now occupy the entire box.

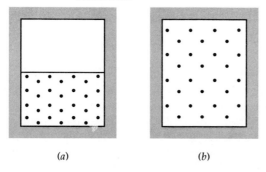

(*a*)   (*b*)

return the gas to its initial state (Fig. II–3a). Such a process will require outside forces. A mass M is placed on a piston at the top of the box (Fig. II–4a). If the mass M is chosen properly, it will descend and compress the gas to exactly its original volume (Fig. II–4b). Now, however, the energy of the gas has increased. By the conservation of energy, the gravitational energy released by the descent of the mass must appear somewhere, and if we assume that the falling piston has no contact with the outside world, then the released energy must appear in the heat energy of the gas, raising its temperature. From Eq. (I–8), the released gravitational energy is $Mg\,\Delta h$, which is the energy increase of the gas. To restore the gas to its initial energy, we must extract from it an amount of heat energy equal to $Mg\,\Delta h$ (Fig. II–4c).

At last, the gas has been returned to its initial condition of volume and energy. However, the *total* system, consisting of the gas and mass, is not in its initial state. The mass has dropped, giving up mechanical energy $Mg\,\Delta h$, and

Figure II–4:   (a) A mass M sits on a piston at the top of a box of gas. (b) The mass falls a distance Δh, pushing down on the piston and compressing the gas. (c) Heat is taken from the compressed gas.

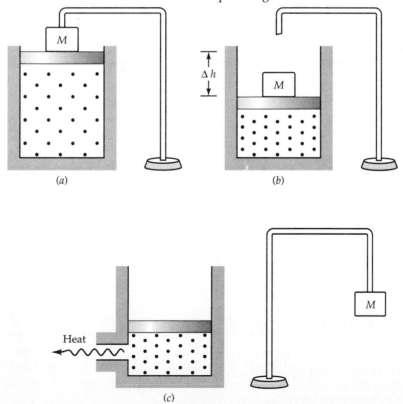

that energy has appeared in the form of heat. Thus, the net effect is that mechanical energy has been converted into heat. An alternative statement of the second law of thermodynamics is that, in isolated systems, mechanical energy is converted into heat energy.

D. THE IRREVERSIBLE FLOW OF HEAT

As mentioned at the beginning of this chapter, the second law of thermodynamics can be stated in several equivalent forms. One form is that an isolated system tends to convert its mechanical energy (bulk motions of few parts) to heat energy (random motions of many parts). In Appendix B–2, we show that this result is associated with the tendency of a system to distribute its energy equally among all of its parts. This tendency may be regarded as an equivalent statement of the second law.

A closely related form of the second law is that two bodies in "thermal contact" with each other—that is, able to exchange heat with each other— evolve to the same temperature. In other words, heat flows from the hotter body to the colder body, cooling off the former and heating up the latter, until the two come to the same temperature. After that, the bodies remain at the same temperature. Since the evolution is such that heat flows only in one direction, from the hotter body to the colder, this evolution is an *irreversible* process. Most statements of the second law of thermodynamics involve irreversible processes. This particular statement—involving heat flow between isolated bodies in thermal contact—resulted from the work of the French scientist Sadi Carnot (1796–1832), the German scientist Rudolph Clausius (1822–1888), and William Thomson (1824–1907). It was the first articulation of the second law.

We can prove the above form of the second law in terms of our probability arguments. For ease of calculation, begin again with an extremely simple system: two boxes of gas in thermal contact, each box containing three molecules of gas. The system is shown in Fig. II–5. The two gases can exchange heat energy in the following way: when a molecule from one of the boxes strikes the wall between them, the molecule can lose energy to the wall (rebounding with less speed than it came in with). This energy can then be gained by a molecule of the other gas striking the other side of the wall (rebounding with more speed than it came in with). In this way, energy can pass from one gas to the other, even though the molecules themselves cannot. The molecules of each gas can also exchange energy among themselves by collisions with each other.

Let us denote the heat energy in gas A by E_A and the heat energy in gas B by E_B. The energies E_A and E_B are, of course, subject to the condition that they add up to the total energy of the system, $E_A + E_B = E_{tot}$, which is fixed (no other forms of energy are involved here).

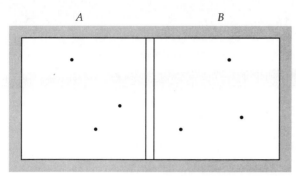

Figure II–5: Two gases, each containing three molecules, separated by a plate that allows heat to pass from one gas to the other.

From Eq. (I–14), recall that the temperature T of a gas is proportional to the average random kinetic energy per molecule, $T = E_H/Nmc$, where E_H is the heat energy of the gas, N is the number of molecules, m is the mass of one molecule, and c is a constant that depends only on the type of gas. The average energy of a molecule is E_H/N. In our situation, $N = 3$ for both gases. To make the notation simpler, but without losing our point, let us assume that $m = 1$ unit of mass and $c = 1$ unit of specific heat. Then the temperatures of gases A and B are

$$T_A = \frac{E_A}{3}, \tag{II–1a}$$

$$T_B = \frac{E_B}{3}. \tag{II–1b}$$

So that we can use some of our previous results for a three-molecule gas, suppose that the system has 11 units of energy, which may be divided between the two gases:

$$E_{tot} = 11 = E_A + E_B. \tag{II–2}$$

What is the probability of various divisions of this total energy between gases A and B? What is the direction of evolution of the system if it does not start in its most probable configuration? Which way does the heat flow?

As a shorthand, let Ω_A stand for the number of states available to gas A, with energy E_A. Similarly, the number of states available to gas B with energy E_B will be called Ω_B. We have already computed the numbers Ω_A and Ω_B for all values of E_A and E_B between 0 and 11. They are in the third column of Table II–3. For example, if $E_A = 7$, then $\Omega_A = 8$, and so on. Now, the number of states available to the *total* system, Ω_{tot}, for a particular division of energy is

$$\Omega_{\text{tot}} = \Omega_A \times \Omega_B. \qquad\qquad \text{(II–3)}$$

This result can be compared to the number of combinations of two dice. Just as the second die may have any of its six possible numbers for each of the six numbers on the first die, gas B may be in any of its Ω_B possible states for each of the Ω_A possible states of gas A, giving a total number of combinations that is the product of Ω_A and Ω_B. The results, derived straight from Table II–3, are shown in Table II–4. There are 302 possible states of the system ($16 + 14 + 24 + 30 + 32 + 35 + 35 + 32 + 30 + 24 + 14 + 16 = 302$). If we wait long enough for energy to be shuffled around among the six molecules, all of these 302 states are equally likely. Then, the probability of a particular configuration (division of energy) is just the fraction of states corresponding to that configuration, shown in the last column of Table II–4.

There are two important results from Table II–4. One is that the most probable configurations are those in which the two gases have the closest temperatures. These are the middle configurations, with one temperature of 5/3 and the other of 2. (Because 11 units cannot be divided into two equal amounts with whole units, it is not possible for the two gases in our simple example to have *exactly* the same temperature.)

The second important result is that the probabilities increase in the direction of transferring heat from the hotter gas to the colder, except at the extremes where one of the gases has zero energy. (These extreme cases become unimportant for larger numbers of molecules.) If the system is started with T_B larger than T_A, the probabilities increase in the direction of increasing E_A and decreasing E_B, and vice versa. Since systems spontaneously evolve

TABLE II–4: States Available to a System of Two Three-Molecule Gases in Thermal Contact with 11 Units of Total Energy, for Various Divisions of the Energy

E_A	E_B	T_A	T_B	Number of states $(\Omega_A \times \Omega_B)$	Probability
0	11	0	11/3	$1 \times 16 = 16$	16/302
1	10	1/3	10/3	$1 \times 14 = 14$	14/302
2	9	2/3	3	$2 \times 12 = 24$	24/302
3	8	1	8/3	$3 \times 10 = 30$	30/302
4	7	4/3	7/3	$4 \times 8 = 32$	32/302
5	6	5/3	2	$5 \times 7 = 35$	35/302
6	5	2	5/3	$7 \times 5 = 35$	35/302
7	4	7/3	4/3	$8 \times 4 = 32$	32/302
8	3	8/3	1	$10 \times 3 = 30$	30/302
9	2	3	2/3	$12 \times 2 = 24$	24/302
10	1	10/3	1/3	$14 \times 1 = 14$	14/302
11	0	11/3	0	$16 \times 1 = 16$	16/302

toward configurations of greater probability, as we have seen, our simple two-gas system will *evolve* in the direction of transferring heat from the hotter gas to the colder, until the two gases have the same temperature.

Notice that one of the Ω's, either Ω_A or Ω_B, always decreases as the system evolves. The gas losing energy during the evolution is actually evolving to a configuration of its own with less probability. However, the gas losing energy is not an isolated system; it is a part of a larger system consisting of both gases, which are interacting with each other through the transfer of heat. What determines the direction of evolution of the *complete* system is the direction of increasing the *total* number of states for the complete system Ω_{tot}.

What we just learned for simple hot and cold gases can be generalized to any hot and cold bodies. Again, the one-way trend of transferring heat from hotter bodies to colder is the essence of irreversibility and of the second law of thermodynamics. Although large departures from this trend are not unlikely for two three-molecule gases, such departures are extremely rare for two realistic gases with many molecules. This is shown in Appendix B.

E. DOING WORK WITH HEAT

1. The Limited Ability to Do Work with Heat and the Universal Decrease of Usable Energy

It is clear that we can convert mechanical energy or work to heat. For example, we can set a pendulum swinging in a box of air. We must do work to push or lift the pendulum; this work is first stored in the pendulum's gravitational energy and then is gradually converted to the heat of the surrounding air as the pendulum slows down. Or we can rapidly rub our hands together and feel them get warm. This is also a process that converts work to heat.

But can we convert heat into work? Yes, but only if there is a temperature difference between a "source" of heat and a "sink" of heat. Furthermore, the process of getting work out of heat in an isolated system always acts to reduce the system's ability to do work. Although the total energy of an isolated system is fixed, the amount of *usable* energy is constantly decreasing. Let's see why.

Figure II-6a shows two gases, one at temperature T_1 and one at temperature T_2. Suppose T_2 is greater than T_1. Then, we will call the gas at temperature T_2 the heat source and that at T_1 the heat sink. Assume that the two gases are thermally connected so that heat can pass between them. Now, from the second law of thermodynamics, we know that heat will flow from the hotter gas to the colder. This flow of heat energy is indicated by the letter H and the arrow in the figure. We can convert some of this heat to work if we place a mass M on a movable piston atop the colder gas. Then that mass will be lifted by the heat energy pouring into the colder gas, just as a balloon expands when

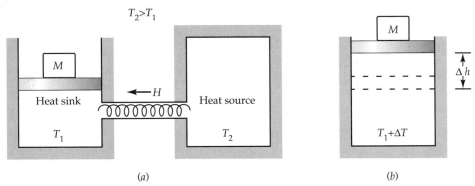

Figure II–6: (*a*) Two gases, or other substances, one at high temperature T_2 and one at low temperature T_1. A mass M sits on a piston on top of the low temperature gas. The two gases are connected, so that heat H can go from the high temperature gas to the low temperature gas. (*b*) The received heat raises the low temperature gas to a temperature $T_1 + \Delta T$ and lifts the mass M a distance Δh.

it is heated. Some of the inflowing heat energy will also go into raising the temperature of the colder gas. Suppose the mass is lifted a height Δh, as shown in Fig. II–6b. Then the conservation of energy, Eq. (I–18b), applied to the colder gas is

$$\Delta E = H - W,$$

or

$$c \, \mathfrak{M} \, \Delta T = H - Mg \, \Delta h, \tag{II–4}$$

where we have substituted $\Delta E = c \, \mathfrak{M} \, \Delta T$ from Eq. (I–15) and $W = Mg \, \Delta h$. Here, H is the added heat energy, c and $\mathfrak{M}$ are the specific heat and mass of the cold gas, respectively, and ΔT is the temperature increase of the cold gas. The work W done by the cold gas is just the increase in gravitational energy of the mass M. (We have changed notation from Eq. (I–15) for the mass of the gas.)

The increased height of the mass represents increased mechanical energy. This energy is *usable* because it can be easily tapped for any purpose. For example, the mass can be removed from the piston and allowed to descend, turning the rotor of an electric generator as it does so.

Notice that heat energy could be used to lift a mass, doing work, only because there was a *temperature difference* between the two gases. If the gases had been at the same temperature initially, then the system would have been in equilibrium and there would have been no *flow* of heat. There might be plenty of heat energy present in both gases, in the form of molecular motions, *but the energy could not be used to do work.* If heat doesn't flow, the weight can't be lifted. The situation of equal temperatures is not unlike coming out of a shower and trying to dry off with a wet towel. If the towel is as wet as you are, you cannot dry off.

Notice also that the transfer of heat from the hotter gas to the colder heats up the colder gas and cools off the hotter body. This transfer reduces the temperature difference between the two gases. As more and more heat is transferred from the hotter gas to the colder, performing more and more work by lifting the weight higher and higher, eventually the two gases will come to the same temperature. At this point, no further work is possible, even though there may still be an enormous amount of heat energy left in the two gases. Thus, heat can be converted to work only for so long, and then the process must come to a halt. *Since heat is continuously flowing from hot bodies to cold bodies everywhere in the universe, as expressed by the second law of thermodynamics, the universe is gradually losing its ability to do work. The total store of usable energy is constantly diminishing.* Not only are all the machines of the universe running down, but the ability to reconvert the resulting heat back to work is lessening with time. There is no way around the one-way street of the second law of thermodynamics.

This startling implication of the second law, which has intrigued and alarmed people since the mid nineteenth century, has been called the *Warmetod*, or "heat death," of the universe. How the second law of thermodynamics applies to the universe as a whole is still debated among physicists.

2. The Efficiency of Machines

Much of the understanding of the second law was motivated by the desire to make machines as efficient as possible. The first such investigations were carried out by the French scientist and engineer Sadi Carnot, in the early nineteenth century, after the industrial revolution was in full swing. In particular, Carnot wanted to know the theoretical maximum efficiency of an engine (or machine) running off heat, a so-called heat engine. From now on, we will use the terms machine and engine interchangeably, a machine simply being a device that can perform work. In his classic memoir "Reflections on the Motive Power of Fire" (1824), Carnot writes,

> Everyone knows that heat can produce motion. That it possesses vast motive-power no one can doubt, in these days when the steam-engine is everywhere so well known. . . . Already the steam-engine works our mines, impels our ships, excavates our ports and our rivers, forges iron, fashions wood, grinds grains, spins and weaves our clothes, transports the heaviest burdens, etc. It appears that it must some day serve as a universal motor, and be substituted for animal power, waterfalls, and air currents. . . . To take away today from England her steam-engines would be to take away at the same time her coal and iron. It would be to

dry up all her sources of wealth, to ruin all on which her prosperity depends, in short, to annihilate that colossal power. The destruction of her navy, which she considers her strongest defence, would perhaps be less fatal. . . . Notwithstanding the work of all kinds done by steam-engines, notwithstanding the satisfactory condition to which they have been brought today, their theory is very little understood. . . . The question has often been raised whether the motive power of heat is unbounded, whether the possible improvements in steam engines have an assignable limit. [*Reflections on the Motive Power of Fire by Sadi Carnot, and other Papers on the Second Law of Thermodynamics*, ed. E. Mendoza (Dover: New York), pp. 3–5]

A heat engine is a generalized version of the steam engine, invented in 1765 by James Watt and in wide use by the late-eighteenth and early nineteenth centuries. In a steam engine, water is converted into steam by burning coal or wood in a boiler. The steam pushes a piston back and forth, turns around the blades of a turbine, or does some other form of work, and then, its energy and pressure exhausted, condenses back into water (Fig. II–7). The process then repeats. Carnot's generalized and abstract heat engine, shown in Fig. II–8, sits between a heat source, or boiler, at high temperature and a heat sink, or condenser, at low temperature. The engine absorbs an amount of heat H_2 from the heat source, performs some work W, such as turning a turbine or lifting a weight, and then deposits whatever heat energy remains, H_1, into the heat sink. Carnot defined the "efficiency" ε of such a heat engine as

$$\varepsilon = \frac{\text{(work done by engine)}}{\text{(heat energy absorbed)}} = \frac{W}{H_2} . \tag{II–5}$$

An efficiency of 1 means that the engine has managed to convert 100% of its absorbed heat energy to work, with no excess heat left over.

Figure II–7: A steam engine.

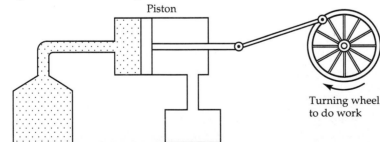

Piston

Turning wheel
to do work

Boiler

Cold
condenser

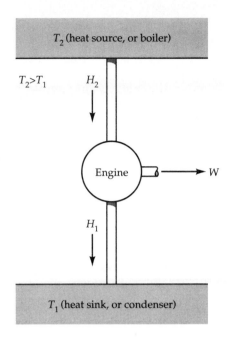

Figure II–8: A heat engine. An amount of heat H_2 comes into the engine from a heat source at temperature T_2. The engine does an amount of work W and deposits the remaining heat energy H_1 into a heat sink at temperature T_1.

Of special interest is a "reversible machine," which is a machine that returns to its initial condition after absorbing heat, performing work, and depositing the leftover heat. A reversible machine is frictionless, meaning that it doesn't waste energy or store heat within itself. All of the heat not turned into work is deposited into the heat sink. Therefore, by conservation of energy, $H_1 = H_2 - W$, or $W = H_2 - H_1$. Substituting this value for W into Eq. (II–5), we see that the efficiency of a reversible machine is

$$\varepsilon = \frac{W}{H_2} = \frac{H_2 - H_1}{H_2} = 1 - \frac{H_1}{H_2}. \tag{II–6}$$

Another property of a reversible machine—and that property giving it its name—is its ability to be run in reverse precisely as if a movie of the machine in operation were played backward. If an amount of work W is *put into* the machine, then a heat H_1 can be taken from the (cold) heat sink, and a heat H_2 can be given to the (hot) heat source. Simply switch the arrows in Fig. II–8.

By contrast, the reverse operation of a nonreversible machine does not look like a movie of its forward operation played backward. Nonreversible machines have friction, and a machine that loses energy in the form of friction when operated in the forward direction will do the same when operated in reverse, just as a block rubbed on a table gets hot whether it is moved forward or backward.

The reverse operation of a heat engine, in which heat is transferred from a cold place to a hot place, doesn't violate the second law of thermodynamics because such an operation requires an input of energy. The engine, heat source, and heat sink do not in this case constitute an *isolated* system; energy (work) is added to the system from the outside. A refrigerator is an example of a heat engine operating in reverse. Heat is continuously taken out of the cold freezer and vented into the warm air outside, but only at the expense of putting energy, in the form of electricity, into the engine.

If two reversible machines with the same efficiency are placed side by side, one operating in the forward direction and one in the reverse direction, then the work produced by the first can be used to run the second. The net effect, after one cycle, is zero. An amount of heat H_2 is taken from the heat source by the first machine but then delivered back again by the second. An amount of heat H_1 is delivered to the heat sink by the first machine but then taken away by the second. And no net work is added to or taken away from the outside world. Thus the machines, heat source, heat sink, and outside world have been returned to their initial condition. The ability to return the total system back to its initial condition, without any net effect, is what characterizes a reversible machine and, in general, a reversible process.

As in Problem II–3 below, Carnot argued that no machine operating between a given heat source and sink could be more efficient than a reversible machine and further that the efficiency of a reversible machine depends only on the temperatures of the heat source and sink.

Problem II–3: Maximally Efficient Machines

Use the second law of thermodynamics to show that no machine can have an efficiency higher than that of a reversible machine operating between the same heat source and sink. [Efficiency is defined by Eq. (II–5)]. Then argue that the efficiency of a reversible machine can depend only on the temperatures of the heat source and sink.

Solution: We will prove the first part by a logical method called "*reductio ad absurdum.*" In this method of proof, an assumption is shown to lead to false consequences, thus disproving the initial assumption. Suppose there exists some machine, which we will call a superefficient machine, that is more efficient than a reversible machine operating between the same heat source and sink. Place a superefficient machine operating in the forward direction alongside a reversible machine operating in the reverse direction, as shown in Fig. II–9. The superefficient machine absorbs an amount of heat H_2 from the heat source, does an amount of work W_*, and delivers the leftover heat H_1 to the heat sink. The reversible machine absorbs an

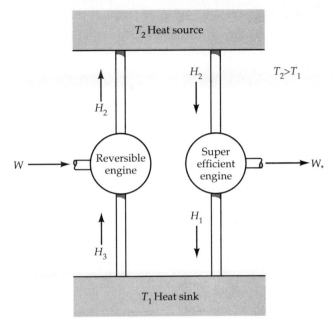

Figure II–9: A reversible heat engine operating in tandem with a superefficient heat engine, both between the same heat source and sink.

amount of heat H_3 from the heat sink, *receives* an input of work W, and delivers heat H_2 to the heat source. Conservation of energy requires

$$W_* = H_2 - H_1,$$
$$W = H_2 - H_3.$$

The efficiency of the superefficient machine is

$$\varepsilon = \frac{W_*}{H_2} = 1 - \frac{H_1}{H_2},$$

and the efficiency of the reversible machine, which we can compute by running it in the forward direction, is

$$\varepsilon = \frac{W}{H_2} = 1 - \frac{H_3}{H_2}.$$

By assumption, the first efficiency is higher than the second, requiring that H_3 be larger than H_1. The net work performed by the two machines is

$$W_* - W = H_3 - H_1,$$

which is positive by the previous result. Thus, the net result for this isolated system of two machines, a heat source, and a heat sink is that an amount of heat $H_3 - H_1$ has been taken from the (cold) heat sink and converted into work. This result violates the second law of thermodynamics, one form of which states that work can be performed by isolated systems only by transferring heat from hot bodies to cold. Therefore, the assumption of the existence of a superefficient machine had to be erroneous from the start. No machine can have an efficiency higher than that of a reversible machine operating between the same heat source and sink. Furthermore, all reversible machines operating between the same heat source and sink must have the same efficiency, because if one had an efficiency higher than the others, it would be a superefficient machine, and we have just proved that such machines do not exist.

Now, for the second part of the problem. We can imagine reversible machines made of a variety of different materials and operating by a variety of different mechanisms. Yet, if they operate between the same heat source and sink, they must have the same efficiency. Since that efficiency cannot depend on the internal details of the machines or on anything to do with the machines, it can depend only on the heat source and the heat sink. The heat source and sink, in turn, can be characterized by their temperatures. Thus, the efficiency of a reversible machine depends only on the temperatures T_2 and T_1 of the heat source and sink.

Rudolph Clausius showed that for a reversible machine, the following relation holds:

$$\frac{H_2}{T_2} = \frac{H_1}{T_1}, \tag{II–7}$$

where the temperature T is in Kelvins. Equation (II–7) may be simply rearranged to give $H_1/H_2 = T_1/T_2$. Substituting this last result into Eq. (II–6), we find that the efficiency of a reversible machine is

$$\varepsilon = 1 - \frac{T_1}{T_2}. \tag{II–8}$$

Note that the efficiency depends only on the temperatures of the heat source and sink, as we expect from Problem II–3.

Since T_2 is greater than T_1, the efficiency ε is always between zero and one, as it must be. A colder heat sink increases the efficiency, as does a hotter heat source. The efficiency is maximized by having the smallest possible ratio of T_1 to T_2. Conversely, if the temperatures of the heat source and sink are equal, $T_2 = T_1$, the efficiency is zero. No work can be produced by a machine operating between bodies of the same temperature. The conversion of heat to work requires a temperature *difference*, as we have discussed before.

Note further that if the heat source and sink are limited in size, as would be the case for any isolated system, then the heat source must unavoidably cool off as it loses heat, and the heat sink must heat up. Thus, as a machine is run through more and more cycles between a heat source and sink, the temperatures T_2 and T_1 approach each other, and the efficiency of the machine gradually decreases until it becomes zero. All these results are manifestations of the second law of thermodynamics.

Problem II–4: Efficiency of Air Conditioners

An air conditioner is a heat engine (machine) working in reverse. Heat is taken from a cool place, like a room, and deposited to a warm place, like the outdoors on a summer day. Energy must be put into the machine to operate it. Analogously to Eq. (II–5), the efficieny ε of a heat engine working in reverse is defined as

$$\varepsilon = \frac{\text{(energy put into machine)}}{\text{(heat energy released)}}.$$

Energy consumption and transfer for air conditioners is usually measured in British thermal units (Btu), where 1 Btu equals 252 calories.

Suppose that the temperature is 21 degrees Celsius (70 degrees Fahrenheit) indoors and 32 degrees Celsius (90 degrees Fahrenheit) outdoors. For every Btu of energy consumed by an air conditioner (in the form of electrical energy taken from the wall plug), what is the maximum number of Btu of heat it can deliver to the outside?

Solution: Whether operating forward or backward, no heat engine can be more efficient than a reversible engine. If the arrows are reversed in Fig. II–8, we obtain the results

$$\frac{W}{H_2} = \frac{H_2 - H_1}{H_2} = 1 - \frac{H_1}{H_2} = 1 - \frac{T_1}{T_2},$$

where we have used the conservation of energy, $W = H_2 - H_1$, and Eq. (II–7). Recall that Eq. (II–7) holds when the temperatures are in Kelvins. The temperatures T_1 and T_2, referring to the cold and hot temperatures, respectively, must be converted into Kelvin: $T_1 = 21\,C + 273\,C = 294K$ and $T_2 = 32\,C + 273\,C = 305$ K. Then we obtain the result that

$$H_2 = \frac{W}{1 - T_1/T_2} = \frac{W}{1 - 294/305} = 27.7W.$$

Thus for every Btu of energy fed into it, the air conditioner can transfer as much as 27.7 Btu of heat to the outside. As the temperature difference between outside and inside gets larger, H_2/W gets smaller, meaning that a given input of energy to the air conditioner can transport less hot air to the outside. This just expresses the fact that it is hard to cool a room when it is hot outside.

Problem II–5: Efficiency of Terrestrial Biological Machines

Many processes on earth, including biological processes, get their energy from the sun and radiate their excess heat into the surrounding air and ultimately into space. The surface of the sun has a temperature of 5000 Kelvins, and space has a temperature of 3 Kelvins. If a terrestrial plant is abstractly represented as a heat engine operating between the sun as a heat source and space as a heat sink, what is the maximum efficiency of such a plant at converting heat to work?

Solution: From Eq. (II–8), the efficiency of a reversible machine operating between the hot sun and cold space is $\varepsilon = 1 - 3/5000 = 0.9994$. Do you think biological processes actually attain this efficiency?

As a digression, we mention that the large difference in temperature between the sun and space plays a vital role in keeping us alive. Considered as heat engines, our bodies must operate between a high-temperature heat source and a low-temperature heat sink. The heat source is provided by our food, whose chemical energy acts as a high-temperature source of energy. In turn, that chemical energy is produced in photosynthesis reactions in plants and derives its high-temperature energy from the high temperature of sunlight (at 5000 Kelvins). Contrary to common belief, we don't need food for energy; we need food for its *high temperature*. Without the second law of thermodynamics, we could simply recycle our bodily energy, converting it back and forth between heat and mechanical energy and never needing to eat a morsel. But the second law decrees that our bodies, or any machines, must inevitably run out of *usable energy* if isolated. To stay alive, we must get a constant infusion of energy at high temperature and release our wasted body heat into something cold. The cold-temperature heat sink is space. The high-temperature heat source can be traced to our food and then to the sun. Lest the red meat eaters gloat and say that they do not need plants to survive, remember that cows eat plants. Either directly or indirectly, the sun keeps us alive. If the sun were the same temperature as space, our body-machines would grind to a halt.

Discussion Questions II–3

What was the motivation for Carnot to work on the theory of heat engines? What are today's analogues of heat engines, and why would it be useful to know their maximum efficiencies?

F. ENTROPY AND ORDER

1. Entropy

The irreversible evolution of systems toward more probable states can be expressed in terms of a quantity called *entropy*, meaning "transformation" in Greek. Rudolph Clausius introduced the word in 1865, although he had worked out the idea a decade earlier. For our purposes, a quantitative definition of entropy is not important and the following qualitative definition will be sufficient: The entropy of a system increases if the number of possible states increases. Entropy may be figured either for an isolated system, or for a system in contact with another system. Since an isolated system always evolves toward configurations of more states (and increasing probability), its entropy increases until the system reaches the configuration with the largest number of states (the most probable configuration). Then its entropy remains constant. Systems that interact with the outside world, that is with other systems, can either gain or lose entropy.

An example will clarify the meaning of entropy. Consider two gases placed into thermal contact with each other, as discussed in section D and shown in Fig. II–10. The two gases can exchange heat with each other, changing each other's number of available states and thus each other's entropy. Suppose that our two-gas system is initially set up so that gas A is hotter than gas B. The complete system, consisting of *both gases*, is isolated; we know that this system will evolve in such a way to increase its *total* number of states, Ω_{tot} = $\Omega_A \times \Omega_B$, Eq. (II–3). From the definition of entropy, the entropy of the complete system will therefore increase, until the two gases have the same temperature.

Let's look now at gas A and gas B individually. During the evolution, energy is transferred from gas A to gas B, since A is hotter. Less energy in A means fewer possible states, as illustrated in Table II–2 for a three-molecule gas. Thus Ω_A decreases. The opposite happens for system B and Ω_B increases. Thus, the entropy of gas A decreases, and that of gas B increases. However, the entropy of the complete system increases during the evolution. Evidently,

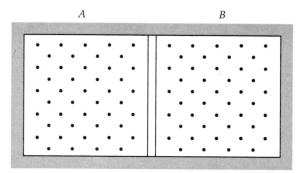

Figure II–10: Two gases, A and B, separated by a plate that allows them to exchange heat with each other.

gas B gains more entropy than gas A loses, so that the net result is an increase in entropy of the complete system. The second law of thermodynamics states that the entropy of any isolated system increases as that system evolves. In an isolated system, entropy can change only in one direction. It gets bigger.

In a paper published in the *Annals of Physics and Chemistry* in 1865, Clausius boldly applied the first and second laws of thermodynamics to the universe as a whole:

> We can express the fundamental laws of the universe which correspond to the two fundamental laws of the mechanical theory of heat in the following simple form: 1. The energy of the universe is constant. 2. The entropy of the universe tends toward a maximum. [*A Source Book in Physics*, ed. William A. Magie (McGraw Hill: New York, 1935), p. 236]

As a student at the University of Berlin, Clausius was at first attracted to history but then settled on a career in physics. Clausius was a brilliant theoretician whose success lay in his ability to generalize his results beyond any particular model. Hence his above application of the second law of thermodynamics to the universe as a whole.

2. Order

Entropy increase is associated with an increase in the *disorder* of a system. We can loosely define disorder as the degree of probability of a configuration. A highly probable configuration is highly disordered, and a highly improbable configuration is highly ordered. Since isolated systems naturally evolve toward configurations of increasing probability, as we have seen, they naturally become more disordered in time.

The above qualitative definition of order and disorder actually accords with our intuition. For example, a box of 5 red marbles and 5 blue ones appears well ordered if all the red marbles are on one side of the box and all the blue on the other. The system appears disordered if the red and blue marbles are mixed together, which is what happens if the box is shaken up.

By counting states, it is easy to show that the configuration in which the marbles are segregated is highly improbable relative to a mixed configuration. A state of the system can be specified by saying whether each marble is a "left-side" marble or a "right-side" marble. Since there are two possible labels for each marble and 10 marbles, there are 2^{10} states of the system, all equally likely. Only two of these states have the red and blue marbles completely segregated (all the red marbles labeled left side and all the blue marbles right side, and vice versa). Thus the probability of a configuration with all the red marbles on one side of the box and the blue on the other is only $2/2^{10} = 1/512$. On the other hand, there are many more states with the marbles mixed together. For example, there are 20 states with one blue marble mixed in with the red ones or vice versa, because starting with either of the two completely segregated states, each of the ten marbles can be crossed over to the other side, for a total of $2 \times 10 = 20$ possible states of this kind. The probabilities of more jumbled configurations are even larger. Low order has high probability.

As another example, consider a swimming pool with a wall in the middle and with ink on one side of the wall and clear water on the other. Remove the wall. What happens? The system begins in a highly ordered and improbable configuration, just as in the previous example with all the red marbles on one side and all the blue on the other. In time, the ink will thoroughly mix with the water, as the system evolves toward a more disordered and more probable configuration. The odds are overwhelmingly stacked against the ink's ever again segregating itself to one side of the swimming pool.

In conclusion, there are many equivalent statements of the second law of thermodynamics, all deriving from our considerations of probability theory: (1) An isolated system naturally evolves toward more probable configurations. (2) An isolated system naturally evolves toward redistributing the available energy equally among its parts. (3) In isolated systems, heat flows from hotter bodies to colder bodies. (4) An isolated system's ability to convert heat into work is constantly decreasing. (5) Although energy is conserved, the *usable* energy in any isolated system is always diminishing. (6) Isolated systems evolve toward increasing entropy. (7) Isolated systems evolve toward decreasing order.

Problem II–6: Entropy and Intelligence

Assume that intelligence is the ability to store and communicate information. Consider a simplified creature with a brain that can store 13 characters, or letters, arranged in a row. When the characters are arranged so that they spell words, they contain information. When the characters don't spell words, no information is present. The characters in our creature's brain are

M, N, B, T, —, O, A, —, I, E, D, G, S. Dashes are word dividers; they indicate the end of one possible word and the beginning of another. Suppose that our creature begins life highly intelligent, with the following arrangement of characters in its brain:

MAN–BITES–DOG

In this arrangement, every character is used in a meaningful word and the words fit together to form a meaningful sentence. This is a highly ordered and special arrangement of the characters. Now, suppose that every year our creature gets bumped on the head and one of the characters in its brain is shaken loose and put back in a random position.

(a) Estimate how long it will take before all the characters have been shaken loose at least once. In this length of time, the characters become completely scrambled.

(b) How many different arrangements of the 13 characters are there?

(c) After all the characters have been shaken loose and replaced in a random order, what is the probability that the original sentence "Man bites dog" will reenter the creature's brain? How many years will it take to return to this original arrangement?

(d) List a few other meaningful sentences that might be spelled. Not all 13 characters must be used in the words of a meaningful sentence. A creature of modest intelligence will employ some characters in meaningful words and have others left over that don't spell anything, as in the character string MAN—DIGS—OBTE. Estimate the number of "partly" intelligent arrangements of characters. Estimate the length of time to form one of these, given the total number of possible arrangements of characters.

Solution: (a) This is an estimation problem. After 13 years, 13 characters will be shaken loose, but some of these will be repeats. We estimate that about a half or a third will be repeats, so about a half or a third of the characters will not have been shaken loose after the first 13 years. After the next 13 years, at least half of the unshaken characters will have been shaken loose. Thus, we might estimate that after about 25 or 30 years, there is high probability (greater than 50%) that all of the characters will have been shaken loose at least once. A similar problem is: How many throws of a die does it take before all 6 faces come up at least once? You can estimate this by tossing a die until all faces come up, writing down the number of tosses required, repeating the experiment several more times, and then averaging the results.

(b) There are 13 positions for the 13 characters, from left to right. The first character can go into any of the 13 possible positions, the second into any of the remaining 12 unoccupied positions, the third into any of the remaining 11 unoccupied positions, and so on, leading to $13 \times 12 \times 11 \times$

$10 \times 9 \times 8 \times 7 \times 6 \times 5 \times 4 \times 3 \times 2 = 6{,}227{,}000{,}000$ possible arrangements. However, all the arrangements that can be obtained by switching the two dashes are identical, since the two dashes are identical. Thus, we have double-counted the number of distinct arrangements, and we must divide by 2. The total number of distinct arrangements of the characters is therefore 3,113,500,000.

(c) MAN—BITES—DOG is only one arrangement out of 3,113,500,000 possible arrangements. After the characters have been completely scrambled, all arrangements are equally likely. Thus the probability of the particular arrangement MAN—BITES—DOG is 1/3,113,500,000. In each successive year, a new arrangement of the characters is obtained. If the probability each year of forming MAN—BITES—DOG is 1/3,113,500,000, it will take about 3 billion years before it is likely that this arrangement will reappear.

(d) An arrangement that meaningfully uses 9 of the characters is DOG—BITES—OBTE. There are $4 \times 3 \times 2 = 24$ arrangements that start out with DOG—BITES—, corresponding to the 24 different arrangements of the leftover characters BEOT. Some of the arrangements are DOG—BITES—OBTE, DOG—BITES—TEOB, etc. A sentence that meaningfully uses only 6 of the characters is I—EAT—MNBGDOS. There are $7 \times 6 \times 5 \times 4 \times 3 \times 2 = 5040$ arrangements that start out with I—EAT—, corresponding to the 5040 different arrangements of the leftover characters BDGMNOS. The shortest possible sentence has the most number of arrangements, since there are the most number of ways of making such a sentence with the leftover characters. If we assume that a marginally intelligent sentence needs at least two words, using up at least 6 characters including dashes, that leaves 7 characters left over, for 5040 different arrangements of such sentences (as in the example above). I will estimate that there are 100 sentences of this variety, leading to $100 \times 5040 = 504{,}000$ possible arrangements. If there are 3,113,500,000 total arrangements, the probability of forming a marginally intelligent arrangement is then estimated to be $504{,}000/3{,}113{,}000{,}000 = 0.00016$. Thus, once the characters are randomized (after the first 25 or 30 years), the number of years needed before even a marginally intelligent sentence reappears is 1/0.00016, or about 6000 years.

G. RESISTANCE TO THE IMPLICATIONS OF THE SECOND LAW

The idea of a static and unchanging universe has had a firm grip on Western thought throughout history. In *On the Heavens*, Aristotle (384–322 B.C.) writes that "the primary body of all is eternal, suffers neither growth nor diminution, but is ageless, unalterable and impassive." Aristotle goes on to conclude that the heavens, being divine and immortal, must be constructed of this primary body, which he called "aether."

The great revolution brought about by Nicolaus Copernicus in 1543, in which the earth was demoted from its position as the fixed center of the universe to a mere planet in orbit about the sun, changed many things, but it did not alter the belief in the constancy and stability of the cosmos. Copernicus explained his own view in *On the Revolutions*: "The condition of being at rest is considered as nobler and more divine than that of change and inconsistency; the latter, therefore, is more suited to the Earth than to the universe."

By contrast, the second law of thermodynamics demands change. Even before the second law was clearly formulated, astute observers were aware of the relentless action of friction. The great British scientist Isaac Newton (1642–1727), who proposed the laws of mechanics and of gravity, wrote in his *Optiks* (1704) that "Motion is much more apt to be lost than got and is always upon the Decay" and that "irregularities" in planetary orbits "will apt to increase, till this System wants a Reformation" by God. The divine "reformation" refers to Newton's devout belief that it was the occasional intervention by God that prevented the world from running down.

Later scientists were not so willing to leave it to God to combat the second law of thermodynamics. A number of leading scientists in the nineteenth and early twentieth centuries rejected or attempted to find a way around the second law of thermodynamics, at least when applied to the universe as a whole. For example, William Rankine (1829–1872), an engineer and seminal theorist on heat engines, proposed that giant reflecting walls in distant space somehow captured the energy lost by decaying systems and refocused it into usable form. In 1862, Lord Kelvin (William Thomson), one of the discoverers of the second law, stated that it was

> impossible to conceive a limit to the extent of matter in the universe; and therefore science points rather to an endless progress . . . than to a single finite mechanism, running down like a clock, and stopping forever.
> [*Popular Lectures and Addresses* (Macmillan: London, 1891), vol. I, pp. 349–350]

In addition to invoking the image of a clock running down, scientists and other intellectuals have often referred to a system in the throes of the second law of thermodynamics as a system approaching a "heat death." As we have seen, energy can be tapped to do work only if there is a temperature difference, and the second law inexorably moves to equalize all temperatures. At this point, a system is incapable of doing work. It has degraded itself to uselessness. It has suffered a heat death.

In 1913, the prominent British geologist Arthur Holmes wrote a little book titled *The Age of the Earth*, in which he summarized and compared all the latest methods for gauging the age of the earth and sun. The last chapter of the book, titled "Thermal Energy of the Sun," notes that the newly discovered atomic energy would be sufficient to maintain the sun's energy supplies for thousands of millions of years. However, even these vast energy stores cannot last forever. Holmes ends the chapter, and book, with the following passage:

> Whence rose this absorption and concentration of energy [in the sun] in the first place? It is evident that once extinct, our sun could not be re-awakened to the warmth of its former activity merely by collision. Gravitational energy alone affords no escape from the ultimate *Warmetod* [heat death], the thermal extinction toward which the universe would appear to be tending. If the development of the universe be everywhere toward the equalisation of the temperature implied by the laws of thermodynamics, the question arises—Why, in the abundance of past time, has this melancholy state not already overtaken us? Either we must believe in a definite beginning, in the creation of a universe furiously ablaze with energy, or else we must assume that the phenomena which we have studied simply reflect our limited experience. Toward the latter alternative we readily incline, the more so because of the hint it affords of cyclic processes in the scheme of Nature. Not only is energy being diffused; somewhere, our hazy conception tells us, energy is being elevated and stored up. With profound insight, Spencer pointed out in 1864 that it is to the attenuated nebulae that we should look for the absorption and concentration of energy. In the universe nothing is lost, and perhaps its perfect mechanism is the solitary and only possible example of perpetual motion. In its cyclic development, we may find the secret of its eternity and discover that the dismal theory of thermal extinction is, after all, but a limited truth. [*The Age of the Earth*, by Arthur Holmes (Harper and Brothers: London, 1913), pp. 120–121]

The phrases "melancholy state" and "dismal theory" in the above excerpt suggests a psychological aversion to the consequences of the second law of thermodynamics: "thermal extinction" of the universe. Scientists frequently reveal their personal attitude toward their subject by the language they use, sometimes even in professional journals.

Discussion Questions II–4

Do you think Arthur Holmes was arguing scientifically when he said that the universe was the only "perpetual motion" machine and was able to replenish itself? Characterize Holmes's world view. How did he reconcile his world view with the second law of thermodynamics?

The Holmes passage provides an opportunity to discuss briefly the diffuse and broad connections between different areas of human thought. The

Spencer that Holmes refers to is British sociologist and philosopher Herbert Spencer (1820–1903). An influential and controversial intellectual figure of the Victorian period, Spencer fiercely believed in individualism and extended this belief to economic free trade, social freedom, and the encouragement of diversity.

Spencer also strongly supported the concept of "progress," a major intellectual theme of the nineteenth century. According to the concept of progress, natural and manmade forces are causing the world to become more and more developed, advanced, organized, and moral with time. This idea was fueled in part by the industrial revolution, which began in England in the previous century and promised a fully mechanized society. On the biological side, the idea of progress was given support by the scientific work of Charles Darwin (1809–1882) and Alfred Russel Wallace (1823–1913) on the theory of evolution. In the theory of evolution, the variation of species within living creatures is explained by the principle of "the survival of the fittest," a phrase coined by Spencer himself [Spencer, *Principles of Biology* (1864), vol. 1, p. 444]. (Darwin called the principle "natural selection.") According to this principle, those animals and plants best adapted to their environment survive to produce offspring and continue their line; creatures not well adapted die before they have babies. Many scientists and nonscientists of the nineteenth century (and some today) interpreted biological evolution as a kind of progress from lower forms to higher, culminating in human beings. Notice that the quote from Lord Kelvin, attacking the slowing-clock metaphor for the universe, says that science points to "endless progress."

The notion of progress—from lower forms to higher, from chaos to organization—was taken up by writers, philosophers, and social thinkers of the nineteenth century. For example, Edward Bellamy's novel *Looking Backward* (1888), set in Boston, describes an ideal social and industrial system of the future. William Morris's *News from Nowhere* (1890) similarly describes an idyllic utopia of great social and ethical progress. Within this general intellectual climate, Spencer had a leading voice. His *Synthetic Philosophy*, a multivolume work completed in 1896, attempted to unify much of human thought and contained volumes on biology, psychology, morality, and sociology. Spencer held that social evolution and progress are carried out by increasing diversity, specialization, and individuation. Increased diversity was accompanied by greater internal order. According to Spencer, society developed from a militant phase, requiring strict governmental control, to an industrial phase, in which social concerns and the rights of the individual eliminated the need for outside control.

The second law of thermodynamics, with its pessimistic forecast of inevitable disorder, inspired dread and animosity in many believers in progress. Spencer and others attempted to find mechanisms to circumvent the second law, and it is one of these mechanisms that Holmes refers to in the passage above. We can now better understand Holmes's thinking. Given the belief in progress that permeated the culture of the nineteenth century, the second law of thermodynamics could not have been a popular idea.

Resistance to the implications of the second law has continued well into the twentieth century. When cosmic rays were discovered in the early twentieth century, no one knew for sure the nature or origin of this powerful radiation from space. The Nobel prize-winning physicist Robert Millikan (1868–1953) proposed an explanation that required an *increase* of order in the cosmos. In particular, Millikan hypothesized that throughout outer space, atoms were continually being formed out of *less organized* particles, and that cosmic rays were produced during the creation of these atoms. In an address in 1928, Millikan stated that

> With the aid of this assumption one would be able to regard the universe as in a steady state now and also to banish forever the nihilistic doctrine of its ultimate heat death. [*Science and the New Civilization*, by Robert Millikan (Scribner's: New York, 1930, pp. 108–109]

Discussion Questions II–5

What did Robert Millikan mean by the "nihilistic doctrine" of the "heat death"? Describe the similarities, if any, in Millikan's and Holmes's world views.

A year after Millikan's address, in 1929, the American astronomer Edwin Hubble found observational evidence that the universe is not in a "steady state" but is expanding, with all the galaxies rushing away from each other. This discovery shattered the longstanding notion of a static universe—a notion assumed as a starting point by Albert Einstein in his cosmological model of 1917. Writing in *The Atlantic Monthly* a few years after Hubble's discovery, the journalist George Gray gave this assessment:

> . . . just as the shifting of bookkeeping accounts into the red measures disintegrating, scattering, dissipating financial resources, so the shifting of starlight into the red [evidence for the expansion of the universe] indicates disintegrating, scattering, dissipating physical resources. It says that the universe is running down, the atomic clock is ticking off suns and planets along with its radioactive particles, matter is dispersing into space and dissipating into radiation, the stupendous pocketbook of the cosmos is emptying itself irrevocably, taking its cash and letting its credit go. . . . According to the generally accepted theory of relativity, space is finite, the universe is a sphere of fixed radius. To entertain this preposterous idea of all the massive star systems racing outward was to accept

a radically new picture of the cosmos—a universe in expansion, a vast bubble blowing, distending, scattering, thinning out into gossamer, losing itself. The snug, tight, stable world of Einstein had room for no such flights. . . . And yet, this is precisely what many of the deep thinkers in science have done. ["Universe in the Red," by George W. Gray, *Atlantic Monthly*, February 1933, pp. 233, 236–237]

Although Gray seems to be discrediting Einstein's entire theory of relativity, in fact, Gray is calling into question only Einstein's own cosmological model, a static model. Other physicists at the time, using Einstein's theory of relativity, found *evolving* cosmological models, solving the same equations and agreeing better with the observation of an expanding universe.

Discussion Questions II–6

Why did George Gray view the expansion of the universe as a process of "dissipation"? What did he mean by saying that an expanding universe was "losing itself"? Discuss the similarities of this language with the language in previous characterizations of the second law of thermodynamics by William Thomson, Arthur Holmes, and Robert Millikan. Discuss the philosophical and psychological associations, if any, evident in Gray's characterization of Einstein's universe. How would you compare Gray's description of the expansion of the universe to the attitudes of Rankine, Kelvin, Holmes, and Millikan about the second law of thermodynamics?

Discussion Questions II–7

How do you feel about the notion of a universe in irreversible dissipation? What are some of the reasons scientists might have resisted the second law of thermodynamics when applied to the universe? Why do you think perpetual motion machines—that is, machines that run forever, without outside energy—have been so appealing throughout history?

H. THE SECOND LAW APPLIED TO HUMAN SOCIETY

The second law of thermodynamics of Carnot, Clausius, and Kelvin, like Darwin's "natural selection" and Dalton's "atom," is a scientific concept that has had impact and application far beyond science. Writers, historians, philosophers, and theologians have all attempted to come to terms with the second law.

In his "Letter to American Teachers of History," in 1910, the historian Henry Adams (1838–1918) applied the second law of thermodynamics to an understanding of human society. Adams, a keen observer of science, prefaces his letter by writing that "few of us are required to look ten, or twenty years, or a whole generation ahead, in order to realize what will then be the relation of history to physics or physiology." Adams then plows ahead with the task himself. Below are some excerpts:

> Towards the middle of the nineteenth century . . . a new school of physicists appeared in Europe . . . who announced a second law of dynamics. (p. 140) The first law said that Energy was never lost; the second said that it was never saved; that, while the sum of energy in the universe might remain constant . . . the higher powers of energy tended always to fall lower, and that this process had no known limit. (p. 141)

> Since the Church had lost its authority, the historian's field had shrunk into narrow limits of rigorously human action; but, strictly within those limits, he was clear that the energy with which history had to deal could not be reduced directly to a mechanical or physico-chemical process. He was therefore obliged either to deny that social energy was an energy at all; or to assert that it was an energy independent of physical laws. Yet how could he deny that social energy was a true form of energy when he had no reason for existence, as professor, except to describe and discuss its acts? . . . therefore he was of necessity a Vitalist, or adherent of the doctrine that Vital Energy [the energy of living creatures] was independent of mechanical law. (p. 146)

Adams is saying here that, in the view of a historian, the workings of human society must involve an energy of some kind, but that energy is not subject to the same physical laws as gravitational energy, kinetic energy, and all the energies physicists speak of.

> At the same moment [middle to late nineteenth century], three contradictory laws of energy were in force, all equally useful to science:—1. The Law of Conservation, that nothing could be added, and nothing lost, in the sum of energy. 2. The Law of Dissipation, that nothing could be added, but that Intensity must always be lost. 3. The Law of Evolution, that Vital Energy could be added, and raised indefinitely in potential, without the smallest apparent compensation. (p. 154)

The Law of Evolution here is Charles Darwin's theory of the causes for the variations in species of living creatures.

> For human purposes, whatever does work is a form of energy, and since historians exist only to recount and sum up the work that society has done, either as State, or as Church, as civil or as military, as intellectual or physical, organisms, they will, if they obey the physical law, hold that society does work by degrading its energies. On the other hand, if the historian follows Haeckel and the evolutionists, he should hold that vital

energy, by raising itself to higher potential, without apparent compensation, has accomplished its work in defiance of both laws of thermodynamics. (p. 156) That the Evolutionist should surrender his conquests seemed quite unlikely, since he felt behind him the whole momentum of popular success and sympathy, and stood as heir-apparent to all the aspirations of mankind. About him were arranged in battalions, like an army, the energies of government, of society, of democracy, of socialism, of nearly all literature and art, as well as hope, and whatever was left of instinct,—all striving to illustrate not the Descent but the Ascent of Man. The *hostis humani generis*, the outlaw and enemy was the Degradationist, who could have no friends, because he proclaimed the steady and fated enfeeblement and extinction of all nature's energies; but that he should abandon his laws seemed a still more preposterous idea. . . . His second law of thermodynamics held its place in every text-book of science. (p. 157)

Here, Adams describes the conflict between the Evolutionists, who think human society is forever progressing upward, and the Degradationists, who believe in the second law of thermodynamics and that all machines, including human societies, are inevitably running down. In the paragraph below, Adams describes signs of the decline of civilization.

Keeping Europe still in view for illustration and assuming for the moment that America does not exist, every reader of the French or German papers knows that not a day passes without producing some uneasy discussion of supposed social decrepitude;—falling off of the birthrate;—decline of rural population;—lowering of army standards;—multiplication of suicides;—increase of insanity or idiocy,—of cancer,—of tuberculosis;—signs of nervous exhaustion,—. . . and so on, without end, coupled with suggestions for correcting these evils such as remind a historian of the Lex Poppeae and the Roman Empire (pp. 186–187)

The battle of Evolution has never been wholly won; the chances at this moment favor the fear that it may yet be wholly lost. The Darwinist no longer talks of Evolution; he uses the word Transformation. The historian of human society has hitherto, as a habit, preferred to write or to lecture on a tacit assumption that human society showed upward progress, even when it emphatically showed the contrary, as was not uncommon; but this passive attitude cannot be held against the physicist who invades his territory and takes the teaching of history out of his hands. Somewhere he will have to make a stand, but he has been already so much weakened by the surrender of his defences that he knows no longer where a stand can be made. As a form of Vital Energy he is convicted of being a Vertebrate, a Mammal, a Monodelphe, a Primate, and must eternally, by his body, be subject to the second law of thermodynamics. Escape there is impossible. Science has shut and barred every known exit. Man can detect no outlet except through the loophole called Mind. (p. 191)

At the Congress of the Italian Society for the Progress of Sciences held at Parma in 1907, Ciamician, the distinguished Professor of the University of Bologna, suggested that the potential of Vital Energy should be taken as the Will. The step seems logical, and to the historian it seems natural. (p. 193) . . . Already the anthropologists have admitted man to be specialized beyond the hope of further variation, so that, as an energy, he must be treated as a weakened Will,—an enfeebled vitality,—a degraded potential. He cannot himself deny that his highest Will-power, whether individual or social, must have proved itself by his highest variation, which was incontrovertibly his act of transforming himself from a hypothetical eocene lemur—whatever such a creature may have been—into a man speaking an elaborately inflected language. . . . The [second] law of thermodynamics must embrace human history in its last as well as in its earliest phase. If the physicist can suggest any plausible way of escaping this demonstration, either logically or by mathematics, he will confer a great benefit on history; but, pending his decision, if the highest Will-power is conceded to have existed first, and if the physicist is to be granted his postulate that height and intensity are equivalent terms, while fall and diffusion are equivalent to degradation, then the intenser energy of Will which showed itself in the primitive extravagance of variation for which Darwin tried so painfully to account by uniformitarian formulas, must have been—and must be now in the constant process of being— degraded and lost, and can never be recovered. (pp. 195–196) [page numbers from *The Degradation of the Democratic Dogma*, by Henry Adams (Harper & Row: New York, 1919); Adams's article is reprinted in *Tendency of History* (Macmillan Company: New York, 1928)]

Discussion Questions II–8

Is Adams's application of the second law to human society metaphorical or literal? If literal, what in human society corresponds to "states of the system?" What corresponds to "configurations of the system?" Is it meaningful to speak of configurations of low and high probability? Is there a sense in which "order" can be defined in human society?

Adams implies that Darwin's "Law of Evolution" involves progress and upward advancement. Do you agree with this interpretation of survival of the fittest?

Adams associates the human Will with the ability for variation and adaptability and argues that the decrease in genetic variation in man, and his weakened Will, are a natural consequence of the second law of thermodynamics. In what sense do you think that decreased genetic variation can be called a decreased Will? Do you think the second law governs either, and if so, how? What is the time period over which the second law would govern human society, according to Adams? Does this time period correspond to the time scale of physical processes associated with the second law? Is Adams optimistic or pessimistic about the future of human society? Do you think that historians today are familiar with the second law of thermodynamics? Should they be?

Discussion Questions II–9

As shown in Appendix B-2, one statement of the second law is that a physical system naturally evolves toward redistributing its energy equally among all of its parts. Zachary Hatch, of the class of 1991 of Princeton University, has proposed that the historical dissolution of empires—in which political power is first consolidated into large, centralized regimes and then ultimately dispersed into many smaller nation-states—can be understood on the basis of the second law of thermodynamics. For example, the end of World War II brought the end to Hirohito's Japan, the modern version of the feudal empire that had existed for hundreds of years. China, too, saw its several-thousand-year-old dynasty fall and be replaced with a socialist state. Very recently, we have witnessed the beginnings of a possible disintegration of the Union of Soviet Socialist Republics. Do you think this application of the second law is justified? If so, what corresponds to the energy of the system? What is the political or social process that corresponds to molecules' bumping into each other and thereby redistributing and sharing the total energy? Does such a process have the needed element of randomness? What corresponds to a state and a configuration of the system? Would the second law, if applicable, preferentially lead to a capitalist or to a socialist society? If true, how would this theory reconcile itself with the *formation* of empires?

Entropy and the second law of thermodynamics figure prominently in the books of the contemporary novelist Thomas Pynchon, among other writers. In Pynchon's *The Crying of Lot 49*, written in the 1960s, a woman named Oedipa Maas wanders around California, experiencing the disillusionment and rebellion of the period. Stanley Koteks, a man at a large engineering firm called Yoyodyne, explains to Oedipa that every engineer who works for Yoyodyne must sign away all patent rights to his inventions. Oedipa is surprised that anyone still invents things.

Koteks looked to both sides, then rolled his chair closer. "You know the Nefastis Machine?" Oedipa only widened her eyes. "Well this was invented by John Nefastis, who's up at Berkeley now. John's somebody who still invents things. Here. I have a copy of the patent." From a drawer he produced a Xeroxed wad of papers, showing a box with a sketch of a bearded Victorian on its outside, and coming out of the top two pistons attached to crankshaft and flywheel.

"Who's that with the beard?" asked Oedipa. "James Clerk Maxwell," explained Koteks, "a famous Scotch scientist who had once postulated a tiny intelligence, known as Maxwell's Demon. The Demon could sit in a box among air molecules that were moving at all different random speeds and sort out the fast molecules from the slow ones. Fast molecules have more energy than slow ones. Concentrate enough of them in one place and you have a region of high temperature. You can then use the difference in temperature between this hot region of the box and any cooler region, to drive a heat engine. Since the Demon only sat and sorted, you wouldn't have put any real work into the system. So you would be violating the Second Law of Thermodynamics, getting something for nothing, causing perpetual motion." (pp. 85–86)

Koteks goes on to tell Oedipa that the Nefastis Machine contains a Maxwell Demon. People with the right "gifts" can stare at the machine and concentrate on which of the two cylinders they want the Demon to heat up. The air in that cylinder then expands, pushing it up. Sometime later, Oedipa visits Nefastis. He brings out his machine and describes it to her:

He then began, bewilderingly, to talk about something called entropy. The word bothered him as much as "Trystero" bothered Oedipa. But it was too technical for her. She did gather that there were two distinct kinds of this entropy. One having to do with heat engines, the other to do with communication. The equation for one, back in the '30's, had looked very like the equation for the other. It was a coincidence. The two fields were entirely unconnected, except at one point: Maxwell's Demon. As the Demon sat and sorted his molecules into hot and cold, the system was said to lose entropy. But somehow this loss was offset by the information the Demon gained about what molecules were where.

"Communication is the key," cried Nefastis. "The Demon passes his data on to the sensitive [person], and the sensitive must reply in kind. There are untold billions of molecules in that box. The Demon collects data on each and every one. At some deep psychic level he must get through. The sensitive must receive that staggering set of energies, and feed back something like the same quantity of information. To keep it all cycling. On the secular level all we can see is one piston, hopefully moving. One little movement, against all that massive complex of information, destroyed over and over with each power stroke." (pp. 105–106)

Later, Oedipa spots an old drunk sailor, shaking with grief and sickness. She helps carry him up to a rooming house, where he sprawls on a mattress.

She remembered John Nefastis, talking about his Machine, and massive destructions of information. So when this mattress flared up around the sailor, in his Viking's funeral: the stored, coded years of uselessness, early death, self-harrowing, the sure decay of hope, the set of all men who had slept on it, whatever their lives had been, would truly cease to be, forever, when the mattress burned. She stared at it in wonder. It was as if she had just discovered the irreversible process. It astonished her to think that so much could be lost, even the quantity of hallucination belonging just to the sailor that the world would bear no further trace of. (p. 128) [*The Crying of Lot 49* by Thomas Pynchon (J. B. Lippincott Company: New York, 1966); page numbers after passages from this edition]

Discussion Questions II–10

How has Pynchon used the second law of thermodynamics? According to the second law, would the entropy of Maxwell's Demon increase or decrease as the Demon sorted the molecules? You can assume that the Demon plus the molecules constitute an isolated system. How would you compare Oedipa's feelings about the loss of information to the scientists' attitudes about the second law of thermodynamics, described in section G?

Before ending this section, I want to emphasize that we should not judge literary and artistic works on the basis of whether they include science or even whether any scientific references are factually correct. Novels and paintings are creations that have value in themselves and define their own realm of truth. When we discuss particular humanistic works here and later on, we will be concerned not with debating the merits of such works but with exploring the wonderful exchange of ideas between science and the humanities.

I. THE SECOND LAW USED TO REFUTE THE THEORY OF EVOLUTION

Henry Morris, president of the Institute for Creation Research, has used the second law of thermodynamics to refute the theory of evolution. He also uses the second law to argue that the universe had to have been created by God. Here are some excerpts from his book *The Troubled Waters* (1981):

The study of biological processes and phenomena indicates that significant evolutionary developments are not observable in the modern world. Similarly the great gaps in the fossil record make it extremely doubtful that any genuine evolution, as distinct from small changes within the kinds, ever took place in the past. . . . There is one weakness in evolutionary theory which goes well beyond the implications of the above difficulties. Not only is there no evidence that evolution ever *has* taken place, but there is also firm evidence that evolution never *could* take place. *The Law of Increasing Entropy* is an impenetrable barrier which no evolutionary mechanism yet suggested has ever been able to overcome. Evolution and entropy are opposing and mutually exclusive concepts. If the entropy principle is really a universal law, then evolution must be impossible. (p. 111)

With God, therefore, is no change. He is eternal and immutable. However, in this present world, everything is under a rule of change. The question is whether the change is up or down, evolution or entropy. (p. 113)

The Second Law proves, *as certainly as science can prove anything whatever*, that the universe had a beginning. Similarly the First Law shows that the universe could not have begun itself. The total quantity of energy in the universe is constant, but the quantity of *available* energy is decreasing. Therefore, as we go *backward* in time, the available energy would have been progressively greater until, finally, we would reach the beginning point, where available energy equalled total energy. Time could go back no farther than this. At this point both energy and time must have come into existence. Since energy could not create itself, the most scientific and logical conclusion to which we could possibly come is that: "In the beginning, God created the heaven and the earth." (pp. 117–118)

Our experience with artificial processes indicates that a code for growth requires an intelligent planner. An architect had to draw the blueprint and a dress designer prepared the pattern. Could mindless, darting particles plan the systematic structure of the elements that they were to form? Even more unbelievably, could these elements later get together and program the genetic code, which could not only direct the formation of complex living systems of all kinds, but even enter into the replication process which would insure the continued production of new representatives of each kind? To imagine such marvels as this is to believe in magic—and magic without even a magician at that! (pp. 125–126)

Does the evolutionist imagine that mutation and natural selection could really perform the function of such an unimaginably complex program? Mutation is not a code but only a random process which, like all random processes, generates disorder in its products. Natural selection is not a code, but only a sort of cybernetic device which snuffs out the disorderly effects of the mutation process. Is the evolutionist really so foolish as to think this kind of mindless interplay could produce the human brain? (pp. 126) . . . One could much more reasonably assume that the sun's energy bathing the stockpiles of bricks and lumber on a construction site will by itself erect an apartment building, an infinitely simpler structural project than the supposed products of organic evolution. (pp. 127) [*The Troubled Waters of Evolution*, by Henry M. Morris (C.L.P. Publishers: San Diego, 1981)]

Discussion Questions II–11

Why do you think Morris used a law from science to refute the theory of evolution? Do you agree with Morris that evolution represents an increase in order? If so, under what circumstances, if any, would this be allowed by the second law of thermodynamics? Is a living creature an isolated system? Is a refrigerator? Do you agree with Morris's arguments? If you disagree, why? Do you agree with Morris that the second law requires a beginning of the universe? Is the biblical statement "In the beginning, God created the heaven and the earth" in contradiction with the theory of evolution? If so, explain how.

Readings

Henry Adams, "A Letter to American Teachers of History," (1910) in *The Degradation of the Democratic Dogma* (Putnam's: New York, 1958).

Sadi Carnot, "Reflections on the Motive Power of Fire," (1824) in *Reflections on the Motive Power of Fire by Sadi Carnot and other Papers on the Second Law of Thermodynamics by E. Clapeyron and R. Clausius*, ed. E. Mendoza (Dover: New York, 1960).

Rudolph Clausius, "The Second Law of Themodynamics," (1850) and "Entropy," (1865) in *A Source Book in Physics*, ed. W. F. Magie (McGraw-Hill: New York, 1935).

Arthur Eddington, "The End of the World: from the Standpoint of Mathematical Physics," (1931) *Nature*, vol. 127, p. 447.

George Gray, "Universe in the Red," (1933) *Atlantic Monthly*, vol. 151, p. 233.

Arthur Holmes, *The Age of the Earth* (Harper and Brothers: London, 1913), chapter VIII.

Henry Morris, "Can Water Run Uphill," in *The Troubled Waters of Evolution* (C.L.P. Publishers: San Diego, 1981).

Thomas Pynchon, *The Crying of Lot 49* (J. B. Lippincott Company: New York, 1966).

ALBERT EINSTEIN

Albert Einstein (1879–1955) was born in Ulm, Germany. His father ran a small electrochemical plant. Einstein did not like the regimentation of school, but he learned much from the mathematics and science books he read on his own. He finished high school in Aarau, Switzerland, and then studied physics and mathematics at the Polytechnic in Zurich.

Unable to get an academic job, Einstein was taken on as an examiner at the Swiss Patent Office in Berne in 1902. During the 7 years he spent at this job. Einstein laid the foundations for much of twentieth-century physics, publishing monumental papers in statistical mechanics (containing ideas similar to those we used in our arguments about the second law of thermodynamics in Chapter II), quantum mechanics (to be discussed in Chapter IV), and special relativity. Each one of these papers helped create a new field of physics. It is an irony of Einstein's career that his work helped build the field of quantum mechanics, even though he himself never could accept the essential uncertainty and indeterminancy required by that subject.

Einstein received his doctorate from the University of Zurich in 1905. Within a few years, he was renowned and had many offers for professorships. After positions at the German University in Prague and at the Polytechnic in Zurich, he became director of the Kaiser Wilhelm Institute for Physics in Berlin in 1914. Here, in 1915, he published his work on the theory of general relativity. After World War I, Einstein came under anti-Semitic attacks, which grew worse until his departure in 1932 for the Institute for Advanced Study in Princeton, where he remained until his death in 1955.

Einstein married the physicist Mileva Maric in 1903 and in 1919 divorced her and married his divorced cousin Elsa Einstein Lowenthal. He had two sons by the first marriage.

Throughout his life Einstein had deep convictions about freedom and humanity, but he was detached from the day-to-day world. In a speech in 1918, Einstein said, "I believe with Schopenhauer that one of the strongest motives that leads men to art and science is escape from everyday life with its painful crudity and hopeless dreariness, from the fetters of one's own ever shifting desires." In his scientific philosophy, Einstein had a profound belief in the beauty of nature and in the ability of the human mind to discover the truths of nature. However, in Einstein's view, this discovery could originate not with experiments but rather only as the "free invention of the human mind," after which the mental invention would be tested against experiment and judged accordingly.

Einstein wrote widely on political, philosophical, and educational matters, although his writings never claimed authority beyond his calling as a physicist. In an address on education in Albany, New York, in 1936, he said, "Desire for approval and recognition is a healthy motive; but the desire to be acknowledged as better, stronger, and more intelligent than a fellow being or fellow scholar easily leads to an excessively egoistic psychological adjustment, which may become injurious for the individual and for the community."

CHAPTER 3

The Relativity of Time

Hear the sledges with the bells—
Silver bells!
What a world of merriment their melody foretells!
How they tinkle, tinkle, tinkle,
In the icy air of night!
While the stars, that oversprinkle
All the heavens, seem to twinkle
With a crystalline delight;
Keeping time, time, time,
In a sort of Runic rhyme,
To the tintinnabulation that so musically wells
From the bells, bells, bells, bells,
Bells, bells, bells—
From the jingling and the tinkling of the bells.
　　　　　Edgar Allen Poe, *The Bells* I. (1849)

Nothing is more basic than time. Time is change. Time is mealtime, time is waking and sleeping, time is sunrise and sunset.

Although in our minds time seems to flow at a fidgety rate, we know that there are timekeeping devices outside of our bodies, ticking off seconds at regular intervals. Clocks, wristwatches, church bells—all divide years into months, months into days, days into hours, and hours into seconds, each increment of time marching after the other in perfect succession. And beyond any particular clock, which might sometimes run slow or fast, we have faith in a vast scaffold of time, stretching across the cosmos, laying down the law of time equally for electrons and people: a second is a second is a second. Time is absolute.

In 1905, Albert Einstein proposed that time is not absolute. Einstein claimed that the rate of flow of time depends on the motion of the clock. A second as measured by one clock corresponds to less than a second as measured by another clock in motion with respect to the first. In other words, time is relative to the observer. Astoundingly, this proposal has been confirmed in the laboratory.

Einstein's quantitative theory of time is called *relativity*. Relativity differs from the previous topics we have discussed, and our approach to it will be different. For the conservation of energy, we could do simple experiments with masses and levers to convince ourselves of the existence of a conserved quantity. And we have all had experience with the second law of thermodynamics, whether we knew it by that name or not. Furthermore, the second law of thermodynamics is essentially nothing more than the theory of probability. The second law might have been logically deduced on the basis of pure mathematics, with no experience of the world.

The relativity of time, on the other hand, is not part of our personal experience with the world. Indeed, it violates that experience. The effects of relativity are extremely tiny at the low speeds we are accustomed to and can be spotted only in experiments with very sensitive instruments. In addition, it seems that relativity is not required simply by logic and mathematics, as is the second law of thermodynamics.

Relativity is a nonintuitive property of nature. In fact, almost all twentieth-century physics—including the quantum theory we will study in the next chapter— is nonintuitive. The physics of the twentieth century, unlike the physics that went before, disagrees with common sense.

Accordingly, we will not be able to approach relativity theory through simple experiments or purely mathematical deductions. Instead, we will follow the path that Einstein took. First, we will review the nineteenth-century science that motivated Einstein to propose the theory of relativity. Then, we will describe the consequences of his theory and work out some of those consequences in quantitative detail. We will compare the theory to some observational results, confirming its validity. Along the way, we will get glimpses of Einstein as a person. We will also discuss some of the deep philosophical consequences of relativity, the influence of philosophy on Einstein's thinking, and the cultural impact of relativity. We will begin with a brief description of relativity.

A. RELATIVITY IN BRIEF

Einstein had two theories of relativity: the theory of special relativity, proposed in 1905, and the theory of general relativity, proposed in 1915. The second theory is a theory of gravity. The first, special relativity, concerns the way in which space and time appear to observers moving at constant speeds relative to each other. (In relativity theory, an "observer" is a person with a set of rulers and clocks for making measurements of distance and time.) In this chapter, we won't discuss general relativity at all, and so we will refer to special relativity simply as relativity.

A fundamental idea in relativity is that there is no condition of absolute rest or motion. There is only relative motion. Absolute motion is motion that can be determined and measured without reference to anything outside the object in motion. We frequently say that we are traveling at some speed, say 60 kilometers per hour (about 37 miles per hour). But what we really mean is that we are moving at 60 kilometers per hour *relative to the road*. In fact, the road is attached to the Earth, which is spinning on its axis, and the Earth is also orbiting the sun at yet another speed. The sun, in turn, is orbiting the center of the galaxy, and so on. So what speed are we actually traveling at?

Einstein claimed that only *relative* motion exists, that an observer moving at constant speed cannot do any experiments to discover how fast, in absolute terms, she is moving or whether she is moving at all. Fixed markers in space, against which all other motions can be measured, simply do not exist. An observer can measure her speed only *relative* to another observer or object. She can say that she is moving at 60 kilometers per hour relative to the ground, or 10,000 kilometers per hour relative to the sun, but to talk of her absolute speed is meaningless.

We all have some experience with these ideas. If we are riding in a car moving at constant speed and do some experiments inside our car, such as tossing balls up in the air and watching them fall, we cannot tell that our car is moving, much less how fast it is moving. If absolute motion existed, then we could determine how fast we were moving just with experiments inside the car, without looking outside the window. However, we know that only by looking out the window can we tell we are moving. Upon looking out the window, we detect our motion *relative* to the trees or the road or other cars. Our speedometer measures our speed relative to the road, but we might have a different speed relative to another car. Of course, if our car suddenly accelerates, we are thrown back in our seat and can definitely tell we are moving. In this case, however, we are no longer traveling at constant speed. Relativity applies to constant speeds.

Motion is closely connected to time. Another idea of relativity is that absolute time does not exist. The relative rate of ticking of two clocks depends on their relative speed. If you and I synchronize our watches, I sit home and you go to the moon (or the grocery store) and back, your watch will read less than mine when you return. Less time will have elapsed for you. You will have aged less than me. This strange result of relativity has been experimentally confirmed.

At the time he formulated his theory of relativity, Einstein was 26 years old and a clerk in a patent office in Bern, Switzerland. To claim that time is not what we think it is—that time does not flow at an absolute rate—took enormous courage, self-confidence, and free thinking. At a young age, Einstein became suspicious of authority of any kind, reacting against unpleasant experiences with religious and political authority. This suspicion of established wisdom may have allowed him to make bold leaps where others could not. As he describes himself in his autobiography:

> Even when I was a fairly precocious young man the nothingness of the hopes and strivings which chases most men restlessly through life came to my consciousness with considerable vitality. Moreover, I soon discovered the cruelty of that chase, which in those years was much more carefully covered up by hypocrisy and glittering words than is the case today. By the mere existence of his stomach everyone was condemned to participate in that chase. Moreover, it was possible to satisfy the stomach by such participation, but not man in so far as he is a thinking and feeling being. As the first way out there was religion, which is implanted into every child by way of the traditional education-machine. Thus I came—despite the fact that I was the son of entirely irreligious (Jewish) parents—to a deep religiosity, which, however, found an abrupt ending at the age of 12. Through the reading of popular scientific books I soon reached the conviction that much in the stories of the Bible could not be true. The consequence was a positively fanatic freethinking coupled with the impression that youth is intentionally being deceived by the state through lies; it was a crushing impression. Suspicion against every kind of authority grew out of this experience, a skeptical attitude towards the convictions which were alive in any specific social environment—an attitude which has never again left me. . . . [Albert Einstein, "Autobiographical Notes" in *Albert Einstein: Philosopher-Scientist* ed. P. A. Schilpp (The Library of Living Philosophers, Evanston, Illinois, 1949), pp. 3–5.]

Discussion Questions III–1

What does the above passage reveal about Einstein as a person? What was the "chase" he was referring to? Can you tell how Einstein felt about authority and established wisdom from this passage? Do you think Einstein's social experiences as a child and young man could have played a role in his proposal of the theory of relativity? If so, how?

B. SCIENCE LEADING TO THE THEORY OF RELATIVITY

When Einstein formulated the theory of relativity, he was trying to make sense of a number of other properties of nature, observed over a long period of time. We will now review those results, in order to understand the logic of his thought.

1. The Relativity of Mechanics

The branch of physics concerned with how masses respond to forces is called mechanics. The great British physicist Isaac Newton (1642–1727) developed the science of mechanics in the seventeenth century, although the Italian physicist Galileo Galilei (1564–1642) and others made important contributions to mechanics before Newton.

The laws of mechanics embody a principle of relativity: a mass acted on by a force responds the same whether the mass is at rest or is moving at constant speed. Suppose we are back in the car, traveling at a constant speed, and we monitor the motion of an object inside the car—say a ball being whirled around on a string. According to the relativity of mechanics, we cannot tell from the ball's motion whether the car is moving or not. For a given force, the ball responds exactly the same to that force no matter what the speed of the car is. For example, suppose that when the car is at rest, a certain tension in a string of 10 centimeters causes the ball to move around in a circle at a speed we measure as 200 centimeters per second. Then, when the car is moving at 60 kilometers per hour (relative to the road) or at any other constant speed, the same tension in the same string will also cause the ball to move around in a circle at a speed of 200 centimeters per second. Since we don't see any change in the behavior of the ball, we cannot possibly measure the car's speed by this means, or even determine that the car is in motion. The principle of relativity in mechanics says that no *mechanical* experiments of any kind can reveal the motion of an observer. An observer can measure his motion only *relative* to another observer or object, but he cannot say that he is moving at such and such a speed, in absolute terms.

The principle of relativity in mechanics was well established both experimentally and theoretically in Newton's day. What Einstein did, 250 years later, was to extend the principle of relativity from mechanics to *all* physics, postulating that *no experiments whatever* can determine an observer's absolute motion. As a consequence, absolute motion does not exist.

2. The Relativity of Electricity and Magnetism

Electricity is a phenomenon associated with positively and negatively charged bits of matter, either at rest or in motion. When the charges are at rest (relative to the observer), the electricity is called static electricity. Electricity has been known since the ancient Greeks, who noticed that the fossil amber, when rubbed with fur, attracts light objects such as feathers. This attraction is a result of static electricity. Electrically charged particles in motion are called current. The electricity that comes out of a wall plug is a current.

Magnetism has also been known since the ancient Greeks and Chinese. Most familiar to us is the deflection of a compass needle by the earth's magnetic field. Magnetism is the force by which certain substances, especially iron, attract similar substances. Magnets are surrounded with magnetic energy, called a magnetic field. You can do a simple experiment to picture the magnetic field around a magnet. Hold a horseshoe-shaped magnet beneath a piece of paper, with the two poles perpendicular to the paper and touching it from below; then sprinkle iron filings on top of the paper. You will see a curved pattern in the location of the iron filings, resembling that shown in Fig. III-1. That curved pattern indicates the shape and direction of the magnetic forces.

In the early nineteenth century, scientists discovered profound connections between electricity and magnetism. First, an electric current in a wire produces a magnetic field around it, as shown in Fig. III–2a. Second, the reverse is also true. A magnet moving inside a loop of wire produces an electrical current in the wire, as shown in Fig. III–2b. Thus, electricity in motion produces magnetism, and magnetism in motion produces electricity. Since electricity and magnetism are so closely related, they are sometimes called by the single name "electromagnetism."

In the mid-nineteenth century, physicists discovered that electromagnetic phenomena embody a relativity principle, although that principle was not fully appreciated until Einstein. What scientists discovered is that the results of electrical and magnetic experiments depend only on the *relative* motion of the parts of the experiment, not on the absolute motion of any single part by itself. For example, when a magnet is moved downward through a loop of wire at a certain speed relative to the wire, exactly the same electrical current is produced in the wire as when the loop is moved upward around the magnet at the same speed, as illustrated in Fig. III–3. The amount of current in the wire can be measured with a meter of some kind, as depicted in the figure. As long as the *relative speed* of the magnet and the wire loop is the same, the cur-

Figure III–1: Pattern formed by iron filings sprinkled around the two poles of a horseshoe magnet. The magnet stands perpendicular to the paper.

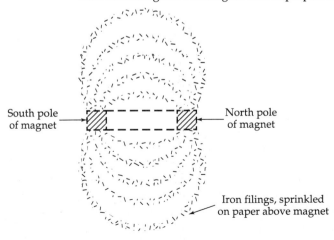

South pole of magnet

North pole of magnet

Iron filings, sprinkled on paper above magnet

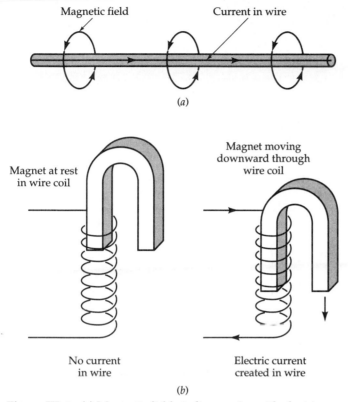

(a)

(b)

Figure III–2: (a) Magnetic field circling a wire with electric current flowing through it. (b) Production of electric current by a moving magnet and wire. When the magnet and wire are both motionless, there is no electric current. While the magnet is moving through the wire, a current flows through the wire.

rent is exactly the same whether the magnet is moving or the wire is moving. Thus, from making measurements with the meter, there is *no way to tell which part of the equipment is moving. There is no state of absolute rest* we can measure against to say whether it is the magnet or the coil that is moving. All that can be measured is relative motion. All we can say is that the magnet and wire coil are moving relative to each other.

Furthermore, if the entire apparatus were placed in a car (or locomotive in the mid-nineteenth century) moving at constant speed, there would be no way to determine the speed of the car from the experiment just described. For if all that can be measured in an electrical or magnetic experiment is the relative motion between its parts, that relative motion remains unchanged when the whole assembly is carried along at some speed. Absolute motion cannot be determined by electrical and magnetic experiments; only relative motion can.

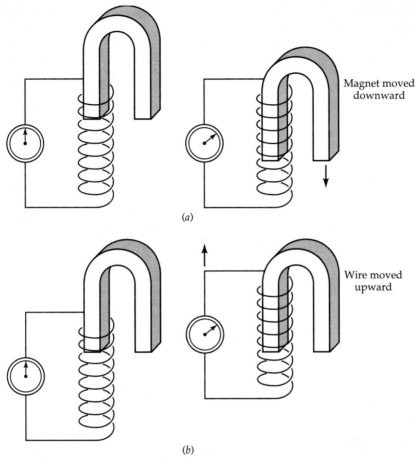

Magnet moved
downward

(a)

Wire moved
upward

(b)

Figure III–3: (a) Electric current produced by a magnet moving down through a wire coil. The amount of current is measured by the current meter. (b) Electric current produced by a wire coil moving upward around a magnet.

All these results for electricity and magnetism found in the nineteenth century seem reminiscent of the principle of relativity for mechanics discovered in the seventeenth century. Evidently, a principle of relativity, forbidding the determination of absolute motion, holds in electromagnetic phenomena as well as in mechanics. Of course, it is much easier to interpret the experimental results with hindsight.

3. The Discovery of Light as an Electromagnetic Phenomenon

In 1865, the Scottish physicist James Clerk Maxwell (1831–1879) showed mathematically that magnets and electric currents should be able to produce traveling waves of electrical and magnetic energy, waves able to move through space on their own, free of the magnets and wires that produced them. First of all, what is a wave? We are familiar with water waves, shown in Fig. III–4. A water wave is an up-and-down variation in height of the surface of a pond or other body of water. A water wave has a crest, where the water height is highest, and a trough, where the height is lowest. The pattern of crest followed by trough moves through the water and constitutes the "traveling wave." The wave could be described by a set of arrows of length proportional to the water height at each location, as shown in Fig. III–4b.

An "electromagnetic wave" is a wave of electrical and magnetic forces traveling through space. Such waves are invisible, unlike water waves. You cannot see electric and magnetic forces, only their consequences. However, we can represent Maxwell's electromagnetic waves by arrows, as we did for water waves. Figure III–5 represents a traveling electromagnetic wave. Here, the vertical arrows are the electrical forces and their lengths are proportional to the magnitudes of those forces. Perpendicular to the electrical forces are the magnetic forces. Figure III–5 is a snapshot of an electromagnetic wave, showing the pattern of electrical and magnetic forces at one instant of time. At a later instant, that pattern will have moved to the right, in the direction of motion of the wave, just as a water wave moves across a body of water.

Figure III–4: (*a*) Water wave traveling on a pond. (*b*) Arrows indicate height of the water at each location of the wave.

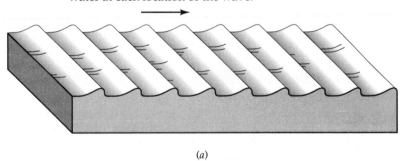

(*a*)

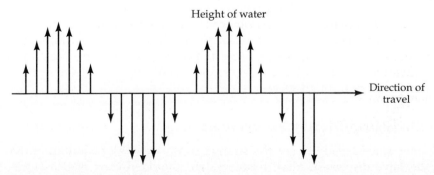

Height of water

Direction of travel

(*b*)

118

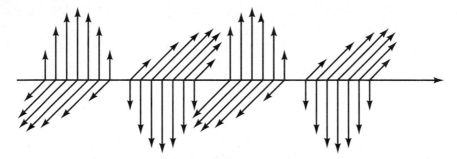

Traveling electromagnetic wave

Figure III–5: Representation of a traveling electromagnetic wave. The vertical arrows indicate the strength and direction of the electric field. The horizontal arrows indicate the strength and direction of the magnetic field. The electric and magnetic fields are perpendicular to each other and both perpendicular to the direction of travel of the entire wave.

Maxwell was a brilliant theoretician. After graduating with high honors in mathematics from Cambridge University in 1854, he began his mathematical investigations of electricity and magnetism. With only pencil and paper, Maxwell discovered in 1865 the theoretical possibility of traveling waves of electromagnetic energy. His calculations required that those waves travel at a particular speed—299,793 kilometers per second, or about 186,000 miles per second. (Recall that a kilometer is 1000 meters, or about 0.62 miles. We will often abbreviate the kilometer as km.) The speed of 299,793 kilometers per second happens to be the same speed at which light travels, a speed first measured in the seventeenth century and well known to Maxwell and other scientists.

To appreciate this discovery, we must realize that nothing about light was used in the derivation of Maxwell's hypothetical waves. The speed of 299,793 kilometers per second that emerged from Maxwell's equations was based entirely on electrical and magnetic phenomena. Yet, there it was, the speed of light. From this result, Maxwell proposed that light, indeed, was a traveling wave of electromagnetic energy.

All of Maxwell's work on electromagnetic waves was theoretical. Then, in 1888, electromagnetic waves were produced in the laboratory, using oscillating electric currents. These waves were found to have the same properties as light. Thus Maxwell's predictions were confirmed. Light is an electromagnetic wave, traveling through space.

4. The Michelson-Morley Experiment

We are now almost ready to see the strange consequences of combining the results we have discussed. First, however, we should review one last crucial experimental result of the nineteenth century, the Michelson-Morley experiment. This experiment, first done in 1887 and considered one of the greatest achievements in physics of all time, attempted to find a state of absolute rest for electromagnetic phenomena. No such state could be found. Some historians of science, notably Gerald Holton of Harvard University, have argued convincingly that Einstein was not particularly influenced by the results of the Michelson-Morley experiment, that many other results, like the kinds we have just discussed, were more influential. In any case, the experiment is extremely important in the history of physics and bears directly on the theory of relativity.

Despite the experiments with magnets and wire coils suggesting that electromagnetic phenomena could not determine absolute motion, late-nineteenth-century physicists continued to believe that such absolute motion must exist. They argued by analogy to water waves and other traveling waves. All known waves required a *material substance* to travel through. Waves could not travel through empty space. A water wave, for example, consists of the up-and-down motions of patches of water. A pattern of such motions moves through the water, and that pattern is the water wave. A sound wave, which carries sound from one place to another, consists of moving molecules bumping into each other. Sound cannot be heard through a vacuum—a fact underscoring the essential role of the material medium in transmitting sound. When sound travels through a room, for example, the air molecules get pushed together, then pulled apart, then together again, and this back-and-forth motion works its way from one end of the room to the other. In fact, this traveling vibration of molecules *is* sound.

For all waves known in the nineteenth century, a material medium, stretching from sender to receiver, was needed to transmit the waves. Thus, as soon as Maxwell and others demonstrated the existence of a new kind of wave, the electromagnetic wave, scientists assumed that these waves too required a material medium for their movement through space. People called that medium the "ether."

The proposed ether would have to pervade all space, since we can detect electromagnetic waves, in the form of visible light, from everywhere, including distant stars. The proposed ether would also have to be very flimsy, since our best attempts to pump all matter out of a jar still allows the transmission of light through the jar, even when sound will not travel through it. Most important, the ether would define a state of absolute rest. You're either at rest with respect to the ether or you're not, and if you're not, you can measure your speed through the ether. Again, scientists reasoned here by analogy to other types of waves. Let's see how the argument goes.

Consider, for example, water waves. Using experiments with water waves, an observer can determine her speed through the water. Let's assume that a water wave travels through water with a speed u. Consider first the case when the observer is at rest in the water. Then, if she splashes and creates water waves, those waves will travel outward from her in all directions *with the same speed u relative to her*, as indicated in Fig. III–6a. She can mea-

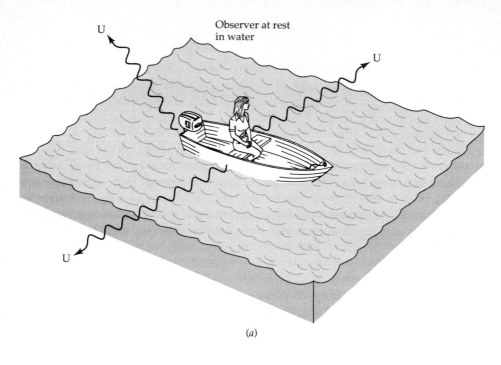

Observer at rest in water

U

U

U

(a)

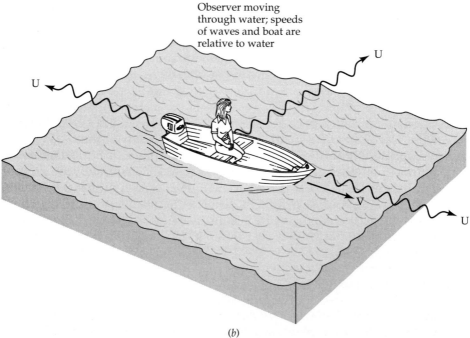

Observer moving through water; speeds of waves and boat are relative to water

U

U

U

v

(b)

Figure III–6: (a) An observer in a boat at rest in a lake. Water waves travel outward at speed u relative to the undisturbed lake. (b) An observer moving at speed v through the lake. Water waves travel outward at speed u relative to the undisturbed lake.

sure the speed of water waves in different directions—for example, with a long ruler attached to her boat and a stopwatch—find that the speed relative to her is the same in all directions, and thus conclude that she is sitting still in the water.

Now, suppose the observer is moving through the water at speed v. She might be paddling or she might have an extremely quiet motor, one so quiet that she cannot tell if it is running. But even so, she can tell she is moving in the water. Because when she creates water waves now, waves in different directions travel at *different speeds relative to her*, as shown in Figs. III–6b and III–6c. Remember that the speed of each wave *relative to the water* is u, and the speed of the boat *relative to the water* is v. A wave traveling in the same direction as the observer's motion through the water will have a speed *relative to her* of $u - v$, and a wave traveling in the opposite direction will have a speed *relative to her* of $u + v$, as shown in Fig. III–6c. Waves traveling in other directions will have relative speeds intermediate between these two extremes.

Not only can our observer determine that she is moving through the water, but she can also determine her direction of motion and speed. Her direction of motion is the direction in which the relative speed of a water wave is *smallest*. Once she has found this direction by experiment, she can then measure the relative speed of a water wave in the opposite direction. If she subtracts the first speed from the second, she gets

$$(u + v) - (u - v) = 2v.$$

Thus, her speed v through the water is $1/2$ the difference of the maximum and minimum relative speeds of a water wave. She can make these determinations of her motion without looking at the shoreline, without looking at the disturbance that her boat makes, and without any knowledge of the mechanism that is propelling her forward. She can determine her speed through the water solely by experiments with water waves. And it is because water waves travel through a material medium and have a definite speed with respect to that medium that such a determination is possible.

In a similar manner, experiments with sound waves allow observers to determine their speed through air, and so on. All waves known to physicists prior to electromagnetic waves required a material medium for their transmission, and experiments with those waves allowed an observer to measure her speed through that medium.

In 1887, the American physicists Albert Michelson and William Morley, working in Cleveland, Ohio, set out to determine the speed of the Earth through the ether by measuring the relative speed of electromagnetic waves traveling in different directions. They began by assuming that the Earth could not be sitting still in the ether: since the ether pervades all space, and since the Earth is constantly changing its direction of motion as it orbits the sun, the Earth could not be permanently at rest with respect to the ether. To make an

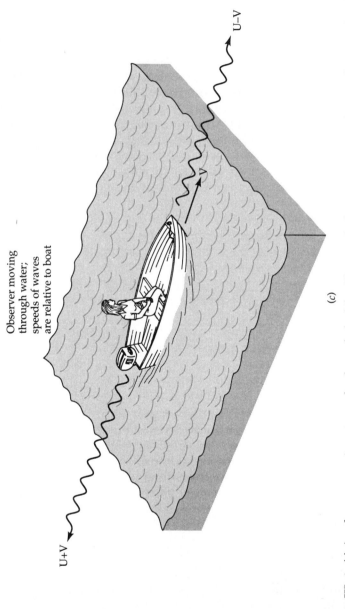

Observer moving
through water;
speeds of waves
are relative to boat

U+V

V

U−V

(c)

Figure III–6: (c) An observer moving at speed v through the lake. The speed of water waves are shown relative to the observer.

analogy with water waves, the Earth is a boat moving through the ethereal fluid. By measuring the speed of electromagnetic waves in different directions, Michelson and Morley would be able to determine the speed of the Earth through the ether (Fig. III–7).

Michelson and Morley built some extremely sensitive and beautiful equipment, which, unfortunately, we will not be able to discuss here, and did their experiment. To their disappointment and shock, they could find no difference in the speed of electromagnetic waves in any direction. In other words, they could not measure the Earth's speed through the ether. Michelson, who was the leader of the experiment, felt like he had failed, even though he was eventually awarded the Nobel prize for his failure, in 1907. He was the first American to win a Nobel in the sciences.

Physicists everywhere were puzzled and disturbed by the null result of Michelson and Morley. What did it mean? Electromagnetic waves, like all other waves, surely must require a medium for their propagation. Why couldn't we measure our speed through that medium?

The answer to these questions came 18 years later, from our young patent clerk. Albert Einstein said that we couldn't measure the Earth's speed through the ether because the ether didn't exist. Light, and other electromagnetic waves, can travel happily through a vacuum. And since a vacuum con-

Figure III–7: The Earth moving through the hypothesized ether at speed v. Observers on the Earth measure the relative speeds of electromagnetic waves emitted in different directions.

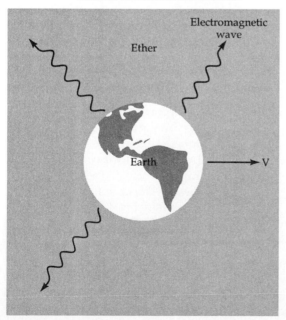

tains no material substance, much less sign posts or fixed points, it is impossible to determine one's speed through a vacuum.

C. THE THEORY OF RELATIVITY

1. Einstein's Postulates

Having carefully studied nineteenth-century physics, Einstein reasoned as follows: If neither electromagnetic phenomena nor mechanical phenomena determine a state of absolute rest, then such a state does not exist. Einstein then made two conjectures, or postulates, from which all the results of his theory of relativity follow. To quote from the first two paragraphs of his landmark paper on relativity, "On the Electrodynamics of Moving Bodies," appearing in the German journal *Annalen der Physik*:

> It is known that Maxwell's electrodynamics—as usually understood at the present time—when applied to moving bodies, leads to asymmetries which do not appear to be inherent in the phenomena. Take, for example, the reciprocal electrodynamic action of a magnet and a conductor [i.e., wire coil]. The observable phenomenon here depends only on the relative motion of the conductor and the magnet, whereas the customary view [a view that includes an ether] draws a sharp distinction between the two cases in which either the one or the other of these bodies is in motion. . . .

> Examples of this sort, together with the unsuccessful attempts to discover any motion of the earth relatively to the "light medium" [ether], suggest that the phenomena of electrodynamics as well as of mechanics possess no properties corresponding to the idea of absolute rest. They suggest rather that . . . the same laws of electrodynamics and optics will be valid for all frames of reference for which the equations of mechanics hold good. We will raise this conjecture (the purport of which will hereafter be called the "Principle of Relativity") to the status of a postulate, and also introduce another postulate . . . namely that light is always propagated in empty space with a definite [relative] velocity c which is independent of the state of motion of the emitting body. [Albert Einstein, "Electrodynamics of Moving Bodies," *Annalen der Physik* vol. 17, pp. 891–892 (1905); translated in *The Principle of Relativity* by H. A. Lorentz, A. Einstein, H. Minkowski, and H. Weyl (New York: Dover, 1953) pp. 37–38.]

Einstein's first postulate says that all observers moving at constant speed (the "frames of reference for which the equations of mechanics hold good") will witness identical laws of physics. In other words, such observers will obtain the same results as each other for experiments involving the same masses, magnets, springs, and so forth. One of the consequences of this postulate is that such observers will not be able to determine their speeds in absolute terms. For, in order to make such a determination, an observer's absolute speed would have to appear in the results of experiments, and, consequently, observers traveling at different speeds would obtain different experimental results. Einstein's first postulate, however, states that all observers moving at constant speeds, even if those speeds are different from each other, must witness the same laws of physics. Absolute speeds do not exist. Fixed markers in space do not exist. This is the principle of relativity.

Einstein's second postulate says, in effect, that the speed of a passing light ray is always measured to be the same by all observers, independent of their own motion. This second postulate is related to the first. Light, as we have discussed, is an electromagnetic phenomenon. And experiments in the nineteenth century showed that electromagnetic phenomena do not allow an observer to determine his absolute speed. If, in contradiction to the second postulate, observers with different speeds *did* measure different relative speeds of a light ray, then they could determine their own speeds through the ether, the frame of absolute rest for electromagnetic phenomena. However, such determinations of absolute motion are impossible according to the relativity principle and the first postulate.

For convenience, we list the two postulates here:

1. Observers in constant motion are completely equivalent. They measure identical laws of physics.
2. The speed of light is the same as measured by all observers, independent of their own speed of motion.

Einstein called his two conjectures "postulates" because he recognized that they were not logically required by the experimental evidence known at the time. They were motivated by the evidence but not required by it. Other physicists, especially the Dutch physicist Hendrik Antoon Lorentz (1853–1928), had very different explanations for the observed phenomena. Lorentz did not do away with the ether or with a state of absolute rest. Instead, he proposed that matter moving through the ether was shrunk by the ether in such a way that the measured speed of light, using shrunken rulers, was always the same, whatever the direction of propagation of the light ray or the motion of the ruler. In this way, light would appear to have the same speed in all directions, independent of the motion of the observer. We will return to these considerations in section E.

2. Consequences of Einstein's Postulates: The Strange Law for Combining Speeds

After beginning with his two postulates, Einstein then deduced their consequences. Indeed, the entire theory of relativity logically follows from the initial two postulates and nothing more. Of course, whether the postulates are true is another matter. Their truth must be established by experiment. We will soon work out for ourselves some of the quantitative results of the theory. First, however, we will just get a feeling for what the postulates mean. Although they may seem harmless at first glance, and even well motivated by the results we have discussed, Einstein's postulates flagrantly violate common sense.

Consider the second postulate: the speed of a passing light ray is measured by all observers to be the same number, 299,793 kilometers per second, *independent of the motions of the observers*. Einstein used the symbol c as an abbreviation for this definite number for the speed of light, that is,

$$c = \text{speed of light} = 299,793 \text{ kilometers per second.} \qquad \text{(III–1)}$$

Make sure you don't confuse this c with the same symbol used to stand for specific heat, discussed in the previous two chapters. Let's see what this postulate means. Suppose there are two observers, one sitting on a bench and another in a train moving by the bench at a speed of 1 kilometer per second, as shown in Fig. III–8. (This is an extremely fast train, but we will not worry about that, since we are conducting a mental experiment.) The observer in the train turns on a flashlight and determines the speed of the emitted light ray: 299,793 kilometers per second. Now, the observer on the bench looks up from her paper and also sees the light ray go by. She measures its speed. On the basis of common sense, we would conclude that if the observer in the train

Figure III–8: An observer on a moving train emits a light ray. Another observer sits on a bench and watches the ray and the train go by.

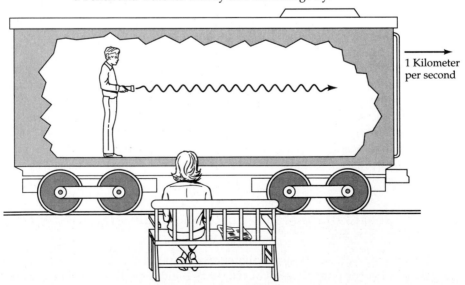

1 Kilometer per second

sees the light ray travel at 299,793 kilometers per second relative to him and the train is moving forward at 1 kilometer per second, then the observer on the bench should clock the light ray's speed as the *sum* of the two speeds:

$$299,793 \text{ km/s} + 1 \text{ km/s} = 299,794 \text{ km/s}.$$

This is what common sense tells us. And it's wrong. According to the second postulate of relativity, the observer on the bench *also* measures the light ray's speed as 299,793 kilometers per second. *The speed of a passing light ray is measured by all observers to be the same number, independent of the motions of the observers.*

We might be tempted to dismiss this bizarre result by allowing it for light rays but not for familiar objects, like balls. However, we can't get off the hook. Suppose that when the observer in the train turns on his flashlight, he also throws a ball forward at a speed extremely close to the speed of light, say 99.9999% the speed of light, so that the ball and the light ray travel along together almost side by side. The observer on the bench will also see the light ray and ball traveling together almost side by side. Thus, that observer will conclude that the ball also has nearly the same speed as the light ray, the same speed as measured by the observer in the train—not a higher speed. The results that apply to the light ray would apply to a ball traveling at practically the speed as light. *Einstein's second postulate really refers to high speeds in general, not just to phenomena involving light.*

How could our common sense be so wrong? Don't relative speeds add together? The young Einstein shrewdly realized that our common sense is based on speeds very, very small compared with the speed of light. A fast automobile, for example, has a speed of about 0.045 kilometers per second, or about $0.045/299,793 = 0.000015\%$ the speed of light. Thus, the effects of relativity are completely unnoticeable in daily life. Our common sense, then, gives approximately the right answers for slow speeds but is grossly wrong for high speeds. We can't have common sense about things we haven't experienced.

Let's see how our common sense can be almost correct for the slow speeds we have experienced but very wrong for high speeds. Suppose that a train is moving at speed v over the ground and that an observer in the train throws a ball forward at speed u_{train} relative to himself. Then our common sense tells us that u_{ground}, the speed of the ball as measured by an observer standing on the ground, is

$$u_{\text{ground}} = u_{\text{train}} + v. \tag{III–2}$$

This is called the law for the combination of speeds. According to common sense, speeds in the same direction combine by just adding them. Indeed, our common sense is almost correct for slow speeds. But if the ball is thrown forward at the speed of light, $u_{\text{train}} = c$, then Eq. (III–2) gives

$$u_{\text{ground}} = c + v,$$

instead of the answer required by Einstein's second postulate (and the answer shown to be correct by subsequent experiments):

$$u_{\text{ground}} = c.$$

The reconciliation of these apparently contradictory results—the first based on common sense and the second on relativity theory—is that the correct law for the combination of speeds is

$$u_{\text{ground}} = \frac{u_{\text{train}} + v}{1 + v \times u_{\text{train}} / c^2}, \tag{III–3}$$

where, for clarity, we have explicitly put in the multiplication sign between v and u_{train}. Equation (III–3) follows mathematically from Einstein's two postulates. In the next section, we will quantitatively work out various results closely related to Eq. (III–3). For the moment, let's just take it as a consequence of the theory and study it.

First of all, Eq. (III–3) gives almost exactly the same result as Eq. (III–2), the common sense result, for all everyday speeds. The reason is that the quantity vu_{train}/c^2 in the denominator of Eq. (III–3) is extremely small when v and u_{train} are ordinary speeds, that is, speeds much smaller than the speed of light c. In such situations, $1 + vu_{\text{train}}/c^2$ is almost equal to 1, and Eqs. (III–3) and (III–2) are almost identical. Suppose, for example, that the train is moving at a speed $v = 0.06$ kilometers per second relative to the ground and the ball is thrown relative to the train with a speed $u_{\text{train}} = 0.04$ kilometers per second (close to the fastest pitch ever made in major league baseball). Then,

$$1 + \frac{vu_{\text{train}}}{c^2} = 1 + \frac{(0.06 \text{ km / s})(0.04 \text{ km / s})}{(299{,}793 \text{ km / s})^2} = 1 + 2.7 \times 10^{-14}$$
$$= 1.000000000000027.$$

Thus, for these speeds, Eq. (III–3) gives

$$u_{\text{ground}} = \frac{0.04 \text{ km / s} + 0.06 \text{ km / s}}{1.000000000000027}$$
$$= 0.0999999999999973 \text{ km / s},$$

which is extremely close to the commonsense result of Eq. (III–2):

$$u_{\text{ground}} = 0.04 \text{ km/s} + 0.06 \text{ km/s} = 0.1 \text{ km/s}.$$

As speeds increase, however, going beyond speeds we are familiar with, the correct law for the combination of speeds, Eq. (III–3), begins to differ more and more from the commonsense law, Eq. (III–2). When the speeds approach the speed of light, the difference becomes large. Suppose, for example, that the train has a speed of half the speed of light, $v = 149{,}897$ kilometers per second, and the speed of the ball relative to the train has the same speed, $u_{\text{train}} = 149{,}897$ kilometers per second. Then the speed of the ball as measured by the observer on the ground is

$$u_{\text{ground}} = \frac{149{,}897 \text{ km} / \text{s} + 149{,}897 \text{ km} / \text{s}}{1 + (149{,}897 \text{ km} / \text{s})(149{,}897 \text{ km} / \text{s}) / (299{,}793 \text{ km} / \text{s})^2}$$

$$= \frac{149{,}897 \text{ km} / \text{s} + 149{,}897 \text{ km} / \text{s}}{1.25} = 239{,}834 \text{ km} / \text{s}.$$

This result differs by about 20% from the commonsense result of Eq. (III–2):

$$u_{\text{ground}} = 149{,}897 \text{ km/s} + 149{,}897 \text{ km/s} = 299{,}793 \text{ km/s}.$$

Notice now another interesting aspect of Eq. (III–3): If $u_{\text{train}} = c$, the speed of light, then $u_{\text{ground}} = c$ also. This can be shown algebraically, without using numbers:

$$u_{\text{ground}} = \frac{c + v}{1 + vc / c^2} = \frac{c(1 + v / c)}{1 + v / c} = c.$$

Thus, *whatever the speed of the train v is,* a ball or light ray moving forward with a speed c as measured by one observer will have the same speed as measured by any other observer, independent of the motions of the observers. This is just what Einstein's second postulate requires. (As we will discuss later, balls can't be thrown with quite the speed of light, but arbitrarily close to it.)
　　We now see that Eq. (III–3), the correct law for the combination of speeds, (1) agrees with the requirements of relativity theory; (2) agrees closely with common sense for slow speeds, upon which common sense is based; (3) provides a smooth connection between combining speeds at slow speeds and at fast speeds. (Slow speeds are speeds small compared to the speed of light.)

3. Consequences of Einstein's Postulates: The Banishment of Absolute Simultaneity

The concept of simultaneity, that is, the occurrence of two or more events at the same moment, is basic to our understanding of time. Common sense tells us that two events that appear simultaneous to one person appear simultaneous to any other person. Common sense tells us that simultaneity is simultaneity, period. Untrue, according to the postulates of relativity. Let's see why. We'll use a mental experiment that Einstein himself considered.

Imagine a train traveling through the countryside. In one of the train cars a person sets up a screen exactly in the center of the car and a light bulb at either end, as shown at the top of Fig. III–9. The screen is wired so that if it is illuminated on both sides at precisely the same instant, a bell rings. Otherwise, the bell remains silent. (Our traveler has brought along some fancy equipment.) The car is darkened, and the person arranges to turn on both light bulbs simultaneously. He can verify that these two events—the turning on of the two light bulbs—happened simultaneously because the bell rings. Since the two light rays traveled the same distance to get to the screen (halfway down the car), traveled with the same speed, and arrived at the center at the same time, they had to have been emitted from their bulbs at the same instant, simultaneously. The situation as seen from the fellow in the train is shown in Fig. III–10a.

Now, let's analyze the situation from the perspective of a person sitting quietly on a bench, shown at the bottom of Fig. III–9. Of course, she also hears the bell ring. The bell either rings or it doesn't. Furthermore, a ringing bell undeniably signals that the two light rays struck the screen at the same time. Prior to this event, however, the sitting observer had to have seen the two light rays emitted *at different times*. The reasoning goes as follows: For her, the train, the screen, and the light bulbs are all in motion. She saw one light ray traveling in the same direction as the train and one in the opposite direction. Thus the first light ray, the trailing light ray, had to travel *more* than half a car

Figure III–9: An observer on a train arranges to have two light bulbs, one at each end of the train, turned on simultaneously. A bell at the center of the train rings if the two light rays strike it simultaneously. Another observer on a bench watches.

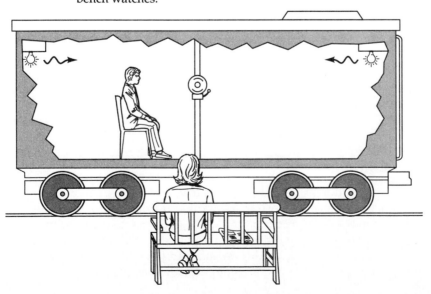

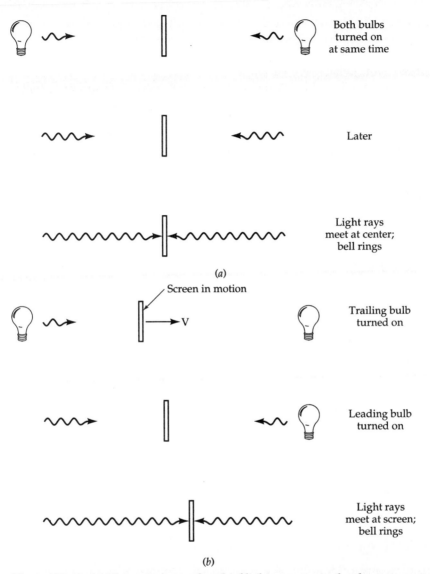

Figure III–10: (*a*) The emission and path of light rays as seen by the person on the
train. (*b*) The emission and path of light rays as seen by the person on
the bench.

length before striking the screen because during its time of transit the train
and screen moved forward a bit. By the same token, the leading light ray traveled *less* than half a car length. The facts are shown in Fig. III–10b. Since both
light rays traveled with the same speed (remember, the second postulate of
relativity requires that the speed of light be the same regardless of any other

motions), arrived at the screen at the same instant, and yet traveled different distances to get there, the trailing ray had to have been emitted *earlier* than the leading ray.

In conclusion, what were simultaneous events to the person in the train were not simultaneous to the person on the bench, contrary to common sense. Simultaneity is not absolute. The reason we *think* that simultaneity is absolute is because the effects of relativity are extremely tiny for everyday speeds, as mentioned before, and our intuition is based on everyday speeds. For example, for a train traveling at 160 kilometers per hour and a car length of 4 meters, the delay between emission of the two bulbs as seen by the person on the bench would be something like one hundredth of one trillionth of a second—hardly noticeable. As you might have guessed, the discrepancies get bigger for bigger relative speeds between the two observers. For a train moving by the bench at almost the speed of light, the trailing light ray would have to travel a very great distance before meeting the screen, while the leading light ray would have to travel only a very short distance. Consequently, the two emission events, as seen from the bench, would be widely separated in time.

If we ponder these disturbing results for a while, we realize that they do not apply only to events involving light rays. In the train illustration, suppose that a baby were born at each end of the train. If each baby were born at the moment the light bulb next to it was turned on, then the observer in the train would say that the babies were born at the same time. However, the observer on the bench would say that the babies were born at different times. Indeed, the emission of light at the two ends of the train could be the herald of *any* two events in those locations, in which case different observers would disagree on whether the two events happened at the same time. Relativity theory deals with *time itself*, not just with the phenomenon of light.

It is just a short step from what we have done to show that the flow of time is not absolute. The relative ticking rates of two clocks depend on their relative motion.

4. Quantitative Derivation of Time Dilation

The relative nature of simultaneity is intimately connected with the flow of time. We will now work out quantitatively what the theory of relativity says about time. All we need are Einstein's two postulates. We will take those postulates as givens and deduce their logical consequences, wherever they lead us. Later, we will have to check the consequences against experiment. *It will be important not to make any assumptions about time and space based on our common sense.* We must resist the temptations of common sense, just as Ulysses tied himself to the mast to resist the deadly songs of the Sirens.

Time is measured by clocks. To understand the flow of time, we need to look carefully at a clock in operation. It doesn't matter what clock we use as long as it is accurate, so we will consider a very simple clock that is easy to analyze. Our clock consists of a rod 1 meter long with a mirror at each end, as shown in Fig. III–11. The bottom mirror has a light detector, which makes a clicking sound every time a light ray hits it. The clock works by bouncing a light ray back and forth between the two mirrors. The light ray reflects off the top mirror down to the bottom mirror, is detected and makes a click, reflects off the bottom mirror to the top mirror and back again, makes another click, and so on. Each click of the light detector is a tick of our clock. Because the light ray travels the same distance between every two reflections (the rod is rigid) and always travels at the same speed, there is an equal interval of time between successive clicks, as in all clocks. So we have constructed a perfectly good clock. With pencil and paper, we will build two identical such clocks. Finally, we will synchronize our two "light clocks" by starting them together.

Figure III–11: A light clock. A light ray bounces back and forth between two parallel mirrors. Every time the ray reaches the bottom mirror, a light detector clicks.

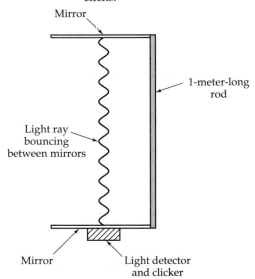

Problem III–1: Synchronization of Clocks

The time of occurrence of an event can be measured only by a clock at the location of that event. Thus, an observer who measures the time of occurrence of events in different locations must have many different clocks, one at each location. Furthermore, these clocks must be synchronized for the time records to have any meaning. To synchronize a group of clocks, the observer must have helpers. Describe how a group of observers, at widely separated locations but lined up in a row, could synchronize their clocks.

Solution: Synchronization means that the clocks are set as if they were all started from zero at the same time. Synchronization is easy if the clocks are all in the same place because then they can simply all be started at the same moment. With separated clocks, the problem is in finding a way for the clocks to communicate with each other. Any method of communication must identify the time and must also take into account that some time is taken for the communciation itself to travel between clocks. The following is one workable scheme: The clocks could be placed along a very long ruler (at rest), with an observer standing beside each clock, and a ray of light could be sent out along the ruler, starting from the zero mark. The observer at the zero mark would start her clock at zero the moment the light ray was emitted. The observer at the 1-kilometer mark would set her clock at

$$\frac{1 \text{ km}}{299{,}793 \text{ km/s}} = \frac{1}{299{,}793} \text{ seconds}$$

at the moment the light ray passed her. This would accurately take into account the time of passage of the light ray from the origin to the 1-km observer. The observer at the 2-kilometer mark would set her clock at

$$\frac{2 \text{ km}}{299{,}793 \text{ km/s}} = \frac{2}{299{,}793} \text{ seconds}$$

the moment the light ray passed her, and so on. This procedure would allow the observers to synchronize their clocks. Considerations of measurement, involving synchronization of clocks among other things, are important in the theory of relativity.

Now, give one of the light clocks we just built to an observer sitting on a bench and a second clock to an observer in a train moving by the bench. Both observers position their clocks so that the rods stand in the vertical direction, perpendicular to the ground and to the direction of the train's motion. Our two clocks are now in place, and we want to analyze their workings from the two observers' points of view.

First, what is the length of the train clock's rod *as measured by the observer on the bench?* We know that the train clock's rod is 1 meter long as measured by the observer in the train, where the rod is at rest, but will it also be 1 meter as measured by the observer on the bench, for whom the rod is in motion?

This may seem like a silly question, but remember that we are exploring territory that may be completely counterintuitive. Thus, we cannot take anything for granted. The length of the moving rod is important because it determines how far apart the mirrors of the moving clock are.

The observer on the bench can measure the length of the train rod by putting two paint brushes on her own rod, one at the bottom, on the 0 mark, and one at the top, on the 1-meter mark. As it moves by her rod, the train rod will get marked by the two paint brushes. The situation is shown schematically in Fig. III–12. Note in the figure that both rods are perpendicular to the direction of relative motion. After the experiment is over and the train has stopped, the observer on the bench can go check the marks on the train rod and see where they fall. If the paint marks on the train rod are separated by less than the length of the train rod, then evidently that rod was longer than 1 meter as measured by the person on the bench. And so on. The observer on the train can measure the length of the bench rod in exactly the same way. He can attach two paint brushes to his rod, one at the top and one at the bottom, and see where the paint marks fall on the bench rod. Both sets of rods and paint brushes are shown in Fig. III–12.

By the first postulate of relativity—that observers moving at constant speed are completely equivalent—the two paint marks on the train rod must be exactly 1 meter apart on that rod, and the same for the bench rod. Let's see why. Analyze the situation first from the point of view of the observer on the bench. Assume, for the sake of argument, that the paint marks on the train's rod fall *less* than 1 meter apart; in other words, assume that the train's rod is longer than 1 meter as measured by the person on the bench. Then the paint

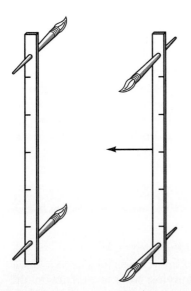

Figure III–12: Two rods parallel to each other and perpendicular to the direction of their relative motion. Each rod has a paint brush sticking sideways from its top and bottom ends, used to mark the other rod as it moves by.

marks on the bench rod will be greater than 1 meter apart, because those marks are made by paint brushes at the top and bottom of the train rod. In particular, if the lower paint mark falls on the 0 mark, the upper paint mark will lie *above* the 1-meter mark on the bench rod. We will see where this assumption leads us.

Consider now the situation as seen by the observer in the train. To the observer in the train, the bench rod is in motion. Since the two observers are completely equivalent, the observer in the train should see the same thing as the observer on the bench. If the observer on the bench concluded that the train rod was longer than 1 meter, the observer in the train must conclude that the bench rod is longer than 1 meter. Thus, his 1-meter paint mark must lie *below* the 1-meter mark on the bench rod. But now we have a contradiction! According to our first analysis, from the perspective of the observer on the bench, the top paint mark on her rod lies above the 1-meter mark; in our second analysis, from the perspective of the person on the train, the top paint mark on the bench rod lies below the 1-meter mark. But there is only one paint mark, and it can't be in two places. Thus our assumption that the train rod appeared longer than 1 meter to the observer on the bench is false. By the same reasoning, the train rod cannot appear shorter than 1 meter to the observer on the bench. In summary, each observer's rod appears exactly 1 meter in length to the other observer, the same length as his own rod.

It has taken a tiresome and subtle argument to prove something that probably appeared obvious at the beginning. Remember, however, that we cannot trust our common sense. Our common sense is based upon only a limited experience with reality. Here, we can allow ourselves only those deductions that are required by the two postulates of relativity. The above argument is not easy. Read it over until you become comfortable with it.

Now we are ready to look at the path of the bouncing light ray in the train's clock from various points of view. As seen by the observer in the train, the train's clock is at rest, as are its two mirrors. The light ray goes straight up and comes straight down between the two mirrors, as shown in Fig. III–13a. (For ease of visualization, we have omitted the rod between the two mirrors.) Let the distance between the two mirrors be D. For the clocks we have built, $D = 1$ meter, but we shall just call the distance D. Let $\Delta t'$ be the time interval between clicks of the train's clock as measured by the observer in the train. Since the light ray travels at speed c and goes a total distance of $2D$ between clicks, the time interval between clicks is

$$\Delta t' = \frac{2D}{c}. \tag{III–4}$$

(Recall that elapsed time is distance traveled divided by speed.)

Let's now look at the path of the light ray in the train's clock *as seen by the observer on the bench*. Because the mirrors are moving for the person on the

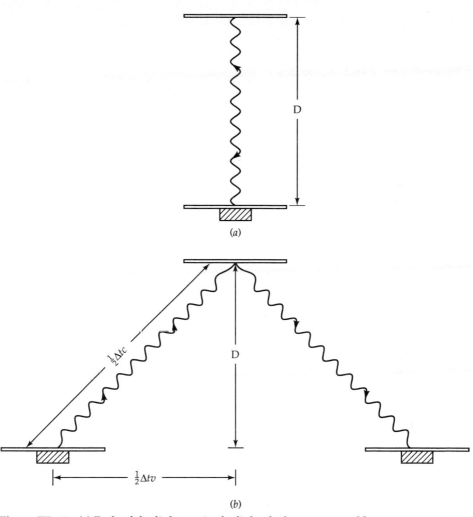

(a)

(b)

Figure III–13: (*a*) Path of the light ray in the light clock as measured by a person at rest with respect to the clock. (*b*) Path of the light ray in the light clock as measured by a person for whom the clock is moving at speed *v*.

bench, the light ray does not go straight up and down. In order to bounce between the two mirrors, the light ray must travel at an angle, as shown in Fig. III–13b. Let Δt be the time interval of one round trip of the light ray as seen by the observer on the bench. This is the time between clicks of the train's clock, as measured by the observer on the bench. Let the speed of the train be v. Then during the time interval for one roundtrip, Δt, the train and each mirror moves horizontally a distance $\Delta t v$. During half of this time interval, $\Delta t/2$, during which the light ray travels from the bottom to the top mirror, both mirrors moves horizontally a distance $(\Delta t/2)v$. This distance is shown in Fig. III–13b. What is the distance the light ray travels in the time $\Delta t/2$, that is, the length of each angled ray in Fig. III–13b? Distance traveled is elapsed time multiplied by speed. According to the second postulate of relativity, the speed of the light ray, c, is the same for the observer on the bench as for the observer in the train. Thus, the light ray travels a distance $(\Delta t/2)c$, as shown in Fig. III–13b. The vertical distance between the two mirrors is the length of the rod between them. We went to great lengths to show that this distance as measured by the observer on the bench is the same as measured by the observer in the train. It is indicated by D in Fig. III–13b, the same D as in Fig. III–13a.

At last we are ready to compute Δt. Notice in Fig. III–13b that the triangle formed by the three lengths $\Delta tc/2$, $\Delta tv/2$, and D is a right triangle, with the length $\Delta tc/2$ being the hypotenuse. (A review of the mathematics of right triangles is given in Appendix A–5.) Thus, we can use the Pythagorean theorem relating the three lengths:

$$\left(\frac{\Delta tc}{2}\right)^2 = \left(\frac{\Delta tv}{2}\right)^2 + D^2.$$

Multiplying both sides of the above equation by 4 and rearranging to put both Δt terms on the left-hand side, we obtain

$$(\Delta t)^2 c^2 - (\Delta t)^2 v^2 = 4D^2,$$

or, combining the Δt terms and dividing by c^2:

$$(\Delta t)^2 (1 - v^2/c^2) = 4D^2/c^2.$$

Finally, taking the square root, we get

$$\Delta t \sqrt{1 - v^2/c^2} = \frac{2D}{c}. \qquad \text{(III–5)}$$

Compare Eqs. (III–5) and (III–4). The left-hand sides are equal to the same quantity, $2D/c$. Thus, the left-hand sides must equal each other:

$$\Delta t \sqrt{1 - v^2/c^2} = \Delta t'.$$

Solving for Δt, we get

$$\Delta t = \frac{\Delta t'}{\sqrt{1 - v^2/c^2}}. \qquad \text{(III–6)}$$

Let's remember what the symbols in Eq. (III–6) mean. The train is moving by the bench at speed v. The speed of light is c = 299,793 kilometers per second. The time between ticks of the train's clock as measured by the person on the bench is Δt. The time between ticks of the train's clock as measured by the person in the train is $\Delta t'$. Equation (III–6) says the two time intervals are not equal! Since $\sqrt{1 - v^2/c^2}$ is always *less* than 1, Δt is always *greater* than $\Delta t'$. In other words, the time between ticks of the train's clock is longer for the person on the bench than for the person in the train, even though they have identical clocks.

Let's put in some numbers. For our particular clocks, D = 1 meter = 0.001 kilometer. Thus, using Eq. (III–4), we have

$$\Delta t' = \frac{2 \times 0.001 \text{ km}}{299,793 \text{ km/s}} = 6.67 \times 10^{-9} \text{ seconds.}$$

This is the time between ticks of the train's clock as measured on the train. From Eq. (III–6), the time between ticks of the train's clock as measured by the person on the bench is

$$\Delta t' = \frac{6.67 \times 10^{-9}}{\sqrt{1 - v^2/c^2}} \text{ seconds.}$$

Every interval between ticks of the train's clock is a factor $1/\sqrt{1 - v^2/c^2}$ longer for the person on the bench than for the person in the train. Every 10 or 1000 ticks is longer by the same factor. Indeed, any interval of time is longer by the same factor because Eq. (III–6) applies to any interval of time. The person on the bench concludes that the moving clock is ticking more slowly than her own *identical* clock, which is at rest from her point of view. This result is called "time dilation." *Moving clocks tick more slowly than clocks at rest.* Of course, whether a clock is at rest or in motion is relative. For the person in the train, the clock in the train is at rest and the clock on the bench is in motion. Therefore, for the person on the train, the clock on the bench ticks more slowly than his clock.

At this point, you might strenuously object. Didn't we argue before that two equivalent observers would measure the same length for two identical rods in relative motion? Now we've deduced that the same two observers measure *different* ticking rates for their two identical clocks. What's going on? If we carefully consider these two situations, we find that they are really not the same. With the meter rods, different measured lengths would have been a *self-contradiction*. Recall that when the rods passed each other, paint marks could be drawn marking the top and bottom of each rod relative to the other rod. Different measured lengths would have led to a contradiction in the location of each paint mark, with each mark lying in one place according to one analysis and in another place according to another analysis. The paint mark can be studied after the experiment, and it obviously can't be in two places at once. For the two clocks, however, the fact that each observer sees the other clock ticking more slowly than his own clock does not lead to a contradiction. A contradiction could arise only if the two clocks could be put together side by side at *two different times*, the first time to synchronize the clocks (for example, to start them both off at noon) and the second to compare later readings. If the two clocks showed different readings the second time they were put together, then there would be a contradiction, analogous to that with the rods. But clocks in constant relative motion cannot be brought together twice. They can be brought together only once, at the moment they pass. Thus no contradiction can arise from the fact that each observer sees the other, moving clock tick more slowly than his own clock. That fact is a consequence of the postulates of relativity. The critical difference between the rods and clocks, then, is this: the lengths of the two rods can be directly compared after bringing them together *only once*, at which time both the top and bottom of each rod can be located relative to the other rod; by contrast, the ticking rates of the two clocks can be directly compared only if they are brought together *twice*, an impossible situation for clocks in constant motion. Each observer can witness the other, moving clock tick more slowly than his own clock without any contradiction.

What we have concluded about the ticking rates of our two clocks applies to *time in general*, not just to the physical mechanism of light clocks or any other clocks. Having constructed our two light clocks, we could now construct two more clocks, of a completely different kind, synchronize each with a light clock, and place each beside a light clock. We could arrange things so that every tick of a light clock corresponds to one tick of the new kind of clock. The tick might be in the form of a swinging arm or a moving ratchet or anything at all. It doesn't matter. Then, our analysis could be repeated. Since each new clock is ticking at the same rate as the light clock sitting next to it, a time dilation for the two light clocks in relative motion would have to occur as well for the two new clocks in relative motion. Indeed, any clock we could possibly construct, as long as it was a good clock, would have to show the same effect. If *all* moving clocks in the train tick slowly compared to clocks at rest on the bench, then we are forced to conclude that time itself is flowing at a slower rate on the train. In fact, as seen by the observer on the bench, all phenomena are occurring more slowly on the train—including our train observer's heart beats, the firings of neurons in his brain, and the chemical reactions in his pancreas. A person on the train is aging more slowly. However, the person on the train experiences nothing abnormal. He cannot tell he is moving by anything happening inside the train, as required by the first postulate of relativity. In fact, he is completely equivalent to the observer on the bench. For him, it is the clocks on the bench that are moving. For him, the clocks on the bench are ticking more slowly than his own clocks. Remember, moving clocks tick more slowly than clocks at rest.

Why haven't we ever noticed the bizarre phenomenon of time dilation—that a clock in motion ticks more slowly than a clock at rest? Because, just as for the odd combination of speeds or the lack of absolute simultaneity, time dilation is extremely small in everyday experience. From Eq. (III–6), the ratio of time intervals of a clock in motion and a clock at rest is the factor $1/\sqrt{1 - v^2/c^2}$. We will call this factor the time dilation factor. A time dilation factor of 1 means no effect whatsoever. For a clock traveling at an everyday speed $v = 0.05$ kilometers per second, the time dilation factor is only

$$\frac{1}{\sqrt{1 - (0.05 \text{ km}/\text{s})^2 / (299{,}793 \text{km}/\text{s})^2}} = \frac{1}{\sqrt{1 - 2.78 \times 10^{-14}}}$$
$$= 1.0000000000000139.$$

Such a tiny difference in the flow of time is completely unnoticeable by humans, but it is measurable by highly sensitive clocks and other equipment. As the relative speed v of a moving clock increases, the time dilation factor becomes larger and larger. Table III–1 shows the time dilation factor for various relative speeds. As you can see, the time dilation effect is very small for ordinary speeds and is less than 10% until v is close to half the speed of light. As v gets closer and closer to c, the time dilation factor gets bigger and bigger. For a speed c, the time dilation factor is infinity. However, as we will see, no material particle, much less a human being, can travel at quite the speed of light, although subatomic particles in gigantic accelerators on earth have been accelerated to 99.99999999% the speed of light! These particle accelerators would not work if they neglected the effects of relativity.

The time dilation formula, and many other results of the theory of relativity, were first worked out by Einstein in his original 1905 paper on relativity.

Problem III–2: Time Dilation

How fast must a moving clock go to tick at half the rate of your stationary clock?

Solution: Use Eq. (III–6), relating the ticking rates of stationary clocks and moving clocks. If the moving clock is ticking half as fast as the clock at rest, then $\Delta t' = \Delta t/2$ for every time interval Δt. Substituting this relation into Eq. (III–6), dividing out the Δt from both sides, and multiplying by $\sqrt{1 - v^2/c^2}$, we get

$$\sqrt{1 - v^2/c^2} = \frac{1}{2}.$$

Squaring this and solving for v^2/c^2, we get

$$\frac{v^2}{c^2} = \frac{3}{4}.$$

Finally, taking the square root, we get

$$\frac{v}{c} = \sqrt{3/4} = 0.866.$$

Thus, the required speed is 86.6% the speed of light.

TABLE III–1: Time Dilation for Various Speeds v
(c is speed of light)

Speed v, km/s	Comment	Time dilation factor, $1/\sqrt{1 - v^2/c^2}$
0	Clock at rest	1
0.05	Fast train	1.000000000000014
0.5	Supersonic jet	1.0000000000014
8	Orbiting satellite	1.00000000035
299.8	0.1% c	1.0000005
29,979	10% c	1.005
149,897	50% c	1.2
269,814	90% c	2.3
296,795	99% c	7.1
299,763	99.99% c	71
299,793	c	Infinity

5. Experimental Tests of Time Dilation: The Disintegration of the Muon

A clear test of the time dilation effect predicted by the theory of relativity is provided by the behavior of subatomic particles called *muons,* which were discovered in 1936. Like many other subatomic particles, muons disintegrate into other subatomic particles soon after they are created. Muons can be produced in nuclear physics experiments in the laboratory, and it has been observed that a muon sitting at rest in the laboratory disintegrates in an average time of 2.2×10^{-6} seconds after its creation. This is called the lifetime of the muon. More precisely, after a period of one lifetime, half of any newly made group of muons have disintegrated.

In addition to their production in the laboratory, muons are created at the top of the Earth's atmosphere. Extremely energetic bombardments from outer space, called cosmic rays, constantly rain down on the Earth and collide with the Earth's upper atmosphere. Such collisions produce muons, which travel downward toward the Earth. How far should one of these muons get before it disintegrates? Even if a muon were traveling at the speed of light, it would seem that a freshly made muon should get only about

$$(299{,}793 \text{ km/s})(2.2 \times 10^{-6} \text{ s}) = 0.66 \text{ km}$$

in its short time to live. Since the top of the Earth's atmosphere is about 10 kilometers above the ground, it seems impossible for the average muon to reach the ground intact. Yet it does. The number of muons striking the top of the Earth's atmosphere can be directly measured, and a large fraction of these do indeed arrive on Earth. How can this be?

The answer is time dilation. For a muon in motion, time is passing more slowly than for a muon at rest. Time is stretched out for the speeding muon. A muon traveling at speed v has a lifetime of $1/\sqrt{1 - v^2/c^2}$ *longer* than its lifetime when at rest [Eq. (III–6)]. Since its lifetime at rest is 2.2×10^{-6} seconds, its lifetime in motion is $2.2 \times 10^{-6}/\sqrt{1 - v^2/c^2}$ seconds. For sufficiently high v, this dilated lifetime is long enough for the muon to reach Earth before disintegrating.

Problem III–3: Long Live the Muon

How fast must the average muon travel in order to reach the Earth from the top of the Earth's atmosphere?

Solution: The average muon at rest lives 2.2×10^{-6} seconds before disintegrating. Analyze everything first from the Earth's point of view. The distance the muon must travel is 10 kilometers. If the muon is traveling at speed v, it will take a time of

$$\Delta t = \frac{10 \text{ km}}{v}$$

to travel to the ground. Let's assume that the muon just barely makes it to Earth and disintegrates as soon as it touches the ground.

Now look at the situation from the point of view of a physicist riding with the muon. For that observer, the muon is at rest, so it has 2.2×10^{-6} seconds to live. We will call this time period $\Delta t'$.

Now back to Earth, where the muon is in motion. The time period $\Delta t'$ is marked by a moving clock, carried along with the muon, and the time period Δt is marked by a clock at rest, sitting on the Earth. According to the theory of relativity, the two time periods are related by Eq. (III–6). Substituting $\Delta t = 10$ kilometers$/v$ and $\Delta t' = 2.2 \times 10^{-6}$ seconds into that equation, we get

$$\frac{10 \text{ km}}{v} = \frac{2.2 \times 10^{-6} \text{ s}}{\sqrt{1 - v^2/c^2}},$$

or, multiplying by $\sqrt{1 - v^2/c^2}$ and dividing by10 km/v,

$$\sqrt{1 - v^2/c^2} = \frac{(2.2 \times 10^{-6} \text{ s}) \, v}{10 \text{ km}}.$$

To solve this equation, divide the denominator and numerator of the right-hand side by 2.2×10^{-6} seconds, getting $v/(4.5 \times 10^6$ km/s). Then divide this last expression by c and multiply by 299,793 km/s (we can always divide and multiply by the same quantity) getting $(v/c)(299{,}793 \text{ km/s})/(4.5 \times 10^6$ km/s) $= 0.067 \, v/c$. Thus, our equation can be written as

$$\sqrt{1 - v^2/c^2} = 0.067 \, v/c.$$

Square both sides and solve for v^2/c^2:

$$\frac{v^2}{c^2} = \frac{1}{1 + (0.067)^2} = \frac{1}{1.0045} = 0.9955.$$

Finally, taking the square root, we obtain

$$\frac{v}{c} = 0.998.$$

Thus, if the muons created by cosmic rays in the upper atmosphere are traveling at a speed of at least 99.8% the speed of light, they will reach the ground before disintegrating, as observed. Without the effects of relativity, we would detect at the ground only a tiny fraction of the muons created in the upper atmosphere.

The detection on the ground of a large fraction of the muons created at the top of the atmosphere is strong confirmation of the theory of relativity. Many other experiments in the twentieth century have further confirmed the predictions of relativity, in quantitative detail. For example, in 1976 physicists

Robert Vessot and Martin Levine of the Smithsonian Astrophysical Observatory launched a highly precise hydrogen-maser atomic clock in a Scout rocket, which traveled up about 10,000 kilometers in altitude at a maximum speed of $v = 9$ kilometers per second, about 0.003% the speed of light. Radio signals allowed the physicists to compare the ticking rate of the moving rocket-borne clock with an identical clock at rest on the ground, and the results showed a time dilation factor in agreement with that predicted by the theory of relativity.

It is impressive to realize that none of these confirming experiments were known when Einstein formulated his theory in 1905. Einstein was not motivated to explain the detection of muons or any particular experimental results. Rather, he wanted to make sense of his general understanding of nineteenth-century physics, in terms of a small number of assumed postulates.

Finally, the example of the muons shows that time dilation is not an *apparent* effect. It is tempting to think that because all observers are equivalent and because each observer sees the other observer's clock ticking slowly, the effects of relativity are only apparent. However, the detection of a muon on Earth is a real event. The average muon either makes it to the ground or it doesn't. It does. And that means it has lived longer in motion than it does while at rest. Relativity, although strange and disturbing, is real.

6. Length Contraction

As we begin to grapple with the results of relativity, turning them over in our minds for logical consistency, we come to another puzzle. Suppose we analyze the average muon's flight from the point of view of a physicist riding with the muon. For her the muon is at rest. It therefore disintegrates after 2.2×10^{-6} seconds, on average. Even if the muon were traveling at 100% the speed of light, it could travel only 0.66 kilometers in its lifetime. Thus, from the point of view of the observer riding along, there should be no way the average muon can survive to reach the ground, 10 kilometers below. Yet it does. The resolution to this apparent contradiction is that the height of the Earth's atmosphere *as measured by the moving observer* is less than 10 kilometers. For the moving observer, the atmosphere has shrunk. This phenomenon is called *length contraction*.

Problem III–4: Moving Lengths

Describe a scheme in which an observer could measure the length of an object in motion.

Solution: It is easy to measure the length of an object at rest. The observer can leisurely place a ruler alongside the object, measuring first the position of one end of the object and then the position of the other end. The situation is different for an object in motion. Here, the two ends of the object must be measured *simultaneously*, since any time lapse would allow the object to move along the ruler. However, as we have seen, simultaneity is relative. Thus, our observer must measure the positions of the two ends of the moving object simultaneously from his perspective. He must have synchronized clocks to do this, so that he knows the two measurements are being done simultaneously.

One scheme would be to hire many helpers along the direction of motion of the moving object, placed at known distances from the first observer. All the helpers first synchronize their clocks with the primary observer, in the manner discussed in Problem III–1. Each helper records the time at which the right end of the object passes him. The first observer records the time at which the left end of the object passes him. He then identifies the particular helper who recorded the same moment of passage as his. From the known position of that helper, our primary observer can figure the length of the object.

We can make length contraction quantitative. Suppose that the height of the Earth's atmosphere, as measured by a person at rest on the Earth, is $\Delta D'$. Suppose further that the average muon just barely makes it to the ground before it disintegrates. Then, we know that its lifetime, as measured by a person on the Earth, is

$$\Delta t = \frac{\Delta D'}{v}.$$

We can write this in terms of the lifetime $\Delta t'$ of the muon as measured by the observer riding with the muon, using Eq. (III–6),

$$\Delta t = \frac{\Delta t'}{\sqrt{1 - v^2 / c^2}} = \frac{\Delta D'}{v}. \qquad \text{(III–7a)}$$

Now let's return to the situation as seen by the physicist riding with the muon. She sees the muon at rest. Imagine a long stick, perpendicular to the Earth's surface and extending from the ground to the top of the atmosphere. The length of this stick is the height of the atmosphere. The observer on the ground sees the stick at rest. But the physicist hurtling toward the ground with the muon sees the stick moving by her at speed v. Let the height of the atmosphere as measured by her, that is, the length of the moving stick, be denoted by ΔD. Since the muon lives a time $\Delta t'$, during which it travels a distance ΔD at speed v, we have

$$\Delta t' = \frac{\Delta D}{v}. \tag{III–7b}$$

Finally, if we multiply Eq. (III–7a) by $\sqrt{1 - v^2/c^2}$ and substitute for $\Delta t'$ from Eq. (III–7b), we get a relation between ΔD, the length of the atmosphere as measured by the observer traveling with the muon, and $\Delta D'$, the length of the atmosphere as measured by the observer at rest on the ground:

$$\Delta D = \sqrt{1 - v^2/c^2}\,\Delta D'. \tag{III–8}$$

Equation (III–8) gives a quantitative expression for how the atmosphere has shrunk from the viewpoint of the physicist traveling with the muon. Note that for this observer, the atmosphere is moving, while for the observer at rest on the ground, the atmosphere is static. Thus, *moving lengths always appear shorter*, by the factor $\sqrt{1 - v^2/c^2}$.

But, you may object again, didn't we show in section 4, with our rods and paint brushes, that lengths don't change under motion? Yes we did, but our analysis was for lengths *perpendicular* to the direction of relative motion. Lengths perpendicular to the direction of motion are indeed measured to be the same by both observers. But lengths *parallel* to the direction of motion, as in the case with the muon and the atmosphere, *are* changed, and that change is given by Eq. (III–8). The difference lies in what can be measured simultaneously by both observers and what cannot. To measure the length of a rod or anything else, you must measure the position of its two ends. The two ends of a meter rod oriented parallel to the direction of relative motion *cannot* be measured simultaneously by both observers. As we remember from the light bulbs at the two ends of the train, two events separated along the direction of motion cannot be simultaneous for both observers. If the measurements of the two ends of a rod parallel to the relative motion happen at the same time according to one observer—constituting two "events"—those two measurements will not have been made simultaneously by another observer. On the other hand, the two ends of a rod oriented perpendicular to the direction of motion *can* be simultaneously measured by both observers. Both observers agree that the tops and bottoms of their rods pass each other at the same moment. This is the important physical difference. *Perpendicular lengths are the same for both observers; parallel lengths are not.*

Sometimes it is confusing to keep track of the quantities with primes and those without primes in Eq. (III–6) and Eq. (III–8). The easiest way to remember is this: The time between ticks of a moving clock is longer, by the factor $1/\sqrt{1 - v^2/c^2}$. The length between marks of a moving stick (oriented along the direction of motion) is shorter, by the factor $\sqrt{1 - v^2/c^2}$. These are the basic results of the theory of relativity. From them, all other results follow.

Problem III–5: Relativistic Squashing of the Enterprise

Suppose the starship *Enterprise*, sitting at rest on the launching pad, is 600 meters tall and 150 meters wide. The ship blasts off, accelerates to cruising speed, and flys by the planet Gork at a speed of 99% the speed of light. By good fortune, a resident of Gork looks up and sees the ship whoosh by. What are the length and width of the ship as measured by the Gorkian? Draw two pictures of the *Enterprise*, one when it is at rest, and one traveling at a speed of 99% the speed of light.

Solution: Lengths perpendicular to the direction of motion are unchanged by the effect of speed, so the width of the *Enterprise* remains 150 meters. Lengths along the direction of motion are compressed, according to Eq. (III–8). Here $\Delta D' = 600$ meters. So,

$$\begin{aligned} \Delta D &= 600 \text{ meters } \times \sqrt{1 - v^2/c^2} \\ &= 600 \text{ meters } \times \sqrt{1 - (0.99)^2} \\ &= 600 \text{ feet } \times 0.14 = 84 \text{ meters}. \end{aligned}$$

A before and after picture of the *Enterprise* is shown in Fig. III–14. Despite the relativistic squashing, the people inside the moving ship don't feel any discomfort at all. For them, the *Enterprise* is at rest and everything is perfectly normal.

Figure III–14: The spaceship *Enterprise* measured when the ship is at rest (left) and when it is traveling at a speed $v = 0.99c$ (right). I hope Trekkie fans will forgive me for simplifying the shape of the *Enterprise*.

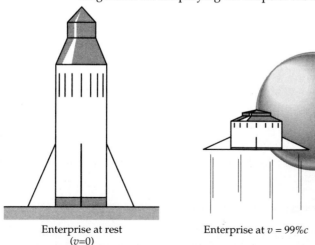

Enterprise at rest
$(v=0)$

Enterprise at $v = 99\%c$

7. Mass-Energy Relation, $E_0 = mc^2$

One result following from length contraction, time dilation, and the first postulate of relativity is the relationship between mass and energy. We will not derive the mass-energy result, as we did the others, but merely state it. If a mass m is traveling with speed v, then the theory of relativity says that its kinetic energy E_K is

$$E_K = \frac{mc^2}{\sqrt{1 - v^2/c^2}} - mc^2. \qquad \text{(III–9a)}$$

Furthermore, the theory of relativity says that the mass, *sitting at rest*, has an energy

$$E_0 = mc^2, \qquad \text{(III–9b)}$$

which is called the "rest energy" of the mass. The total energy E of the mass (not counting gravitational energy, or heat energy) is the sum of the kinetic energy and the rest energy:

$$E = E_K + E_0 = \frac{mc^2}{\sqrt{1 - v^2/c^2}}. \qquad \text{(III–10)}$$

It is the total energy E that enters into the energy conservation laws.

There are a couple of new things here. First, Eq. (III–9a) does not look like the formula for kinetic energy that we derived in Chapter I, Eq. (I–7):

$$E_K = \tfrac{1}{2}\, mv^2. \qquad \text{(III–11)}$$

Second, the rest energy E_0 is a completely new concept. Let's look first at the two different formulas for kinetic energy, Eq. (III–9a) and Eq. (III–11). Almost all the formulas that we derived in Chapter I for various kinds of energy were based on nineteenth-century physics, before the understanding of relativity. We didn't worry about time dilation or length contraction. And, in fact, for almost all types of experiments, where speeds are slow compared with the speed of light, the effects of relativity are tiny, and we are justified in ignoring them.

Let's look at the relativity factor more carefully. If v/c is much smaller than 1, then an excellent *approximation* to the relativity factor $1/\sqrt{1 - v^2/c^2}$ is

$$\frac{1}{\sqrt{1 - v^2/c^2}} \approx 1 + \frac{v^2}{2c^2}, \qquad \text{(111–12)}$$

where the $\approx$ sign means that the equation represents an approximation rather than an exact equality. We can verify that Eq. (III–12) is approximately correct for small values of v/c by trying various values. For example, if $v/c = 0.001$, the exact relativity factor is $1/\sqrt{1 - v^2/c^2} = 1.000000500000375$, to 15 significant figures, while the approximation is $1 + v^2/2c^2 = 1.000000500000000$. The two differ only in the thirteenth decimal place. For $v/c = 0.1$, which is far larger than any speed normally encountered, the exact relativity factor is 1.00504, to five significant figures, while the approximation to the relativity factor is 1.00500. Thus, Eq. (III–12) represents an excellent approximation unless the speed is very close to the speed of light. To see algebraically why Eq. (III–12) works as a good approximation, square both sides and then multiply both sides by $1 - v^2/c^2$:

$$\frac{1}{1 - v^2/c^2} \approx 1 + \frac{v^2}{c^2} + \frac{v^4}{4c^4},$$

$$1 \approx 1 - \frac{3v^4}{4c^4} - \frac{v^6}{4c^6}.$$

Notice that the terms in v^2/c^2 canceled each other out in the last step. This is the key point. If v/c is small, then v^2/c^2 is even smaller, $3v^4/4c^4$ is smaller still, and $v^6/4c^6$ is smaller still. The last step shows that our approximation is equivalent to neglecting the extremely small terms $3v^4/4c^4$ and $v^6/4c^6$.

If we substitute the approximation of Eq. (III–12) into Eq. (III–9a), we obtain

$$E_K \approx mc^2 \left(1 + \frac{v^2}{2c^2}\right) - mc^2 = \frac{1}{2}mv^2, \qquad \text{(III–13a)}$$

and the total energy of the particle is approximately

$$E = E_K + E_0 \approx \tfrac{1}{2}mv^2 + mc^2. \qquad \text{(III–13b)}$$

Aha! Equation (III–13a) looks just like our previous expression for kinetic energy, Eq. (I–7) and Eq. (III–11). So, for all normal speeds, Einstein's formula for kinetic energy, Eq. (III–9a), does indeed agree very closely with the previous formula for kinetic energy. For the less familiar high speeds, however, the two formulas do not agree.

Next comes the rest energy, mc^2. The concept of the rest energy of a mass was not anticipated by Einstein or other scientists. But it emerges unavoidably out of the theory of relativity. A mass m at rest has an *equivalent energy content* of mc^2, which must be included in the mass's total energy in order to make the law of the conservation of energy still hold under the postulates of relativity.

A mass m at rest can disappear completely as long as an energy mc^2 appears in its place. Mass by itself need not be conserved. It is the total energy that must be conserved. Because c is such a huge speed compared to ordinary speeds, mc^2 is usually much bigger than $mv^2/2$. Thus, masses are capable of producing enormously more energy than they would have just by virtue of their kinetic energy.

Problem III–6: Rest Energy

In 1991, the city of Boston and its vicinity had a population of about 3 million people and consumed about 3×10^{15} joules of energy per day. How much mass would it take to provide the daily energy needs of the metropolitan Boston area, assuming that you had a mechanism to convert the mass completely to energy?

Solution: Use Eq. (III–9b), and remember your units. To express energy in joules, mass must be expressed in kilograms and speeds in meters per second. The speed of light is $c = 299{,}793$ kilometers per second $= 2.998 \times 10^5$ kilometers per second $= 2.998 \times 10^8$ meters per second. The required energy from rest energy is

$$E_0 = mc^2 = 3 \times 10^{15} \text{ joules,}$$

or

$$m = \frac{3 \times 10^{15} \text{ joules}}{(2.998 \times 10^8 \text{ m}/\text{s})^2} = 0.033 \text{ kg} = 33 \text{ grams.}$$

This is about the mass of a golf ball! Perhaps now you can understand the enormous energy content of mass and the fame of the formula $E_0 = mc^2$. Of course, humankind has yet to develop any way of converting a mass completely to energy. The best we have done so far, in nuclear fission reactors, is to convert about 0.1% of a mass to energy. Still, even with only 0.1% efficiency, it would take only about 30 kilograms of mass (70 pounds) to power the entire metropolitan area of Boston for a day. By comparison, the burning of fossil fuels—including coal, oil, and gas—requires about 10^8 kilograms, or about 100,000 tons.

---

8. Meaning of the Speed of Light

The speed of light c occurs all over the place in the theory of relativity—in the formula for time dilation, for length contraction, and for the mass-energy relation. What's so special about the speed of light? Isn't it just a number—299,793 kilometers per second?

As you remember, the speed of light came out of the fundamentals of electricity and magnetism. The speed of light is what physicists call a "fundamental constant of nature." It is a basic property of our universe and plays a role in practically every phenomenon. Aliens in another galaxy would be familiar with the speed of light, just as they would be familiar with the mass of an electron or the concept of absolute zero or the fact that space has three dimensions. Any advanced civilization would quickly find out that 299,793 kilometers per second is a fundamental speed in the universe. In fact, it is the cosmic speed limit. As Eq.(III–9a) shows, the kinetic energy of a moving mass mounts to infinity as its speed approaches the speed of light. [As v/c gets closer and closer to 1, the denominator on the right-hand side of Eq. (III–9a) gets closer and closer to 0.] If we think of trying to accelerate a mass all the way up to the speed of light, it would require an *infinite* amount of energy to do so. Therefore, no mass can travel quite as fast as light. Light, and other forms of electromagnetic radiation and pure energy, travel at speed c, but no mass can. All masses must travel at less than the speed of light, as measured by any observer.

Why our universe has a speed of light equal to 299,793 kilometers per second, and not 299,792 kilometers per second or 299,794 kilometers per second, is a mystery. It is conceivable that other universes could exist with other values for the speed of light. But our universe has its particular value, and that value is a fundamental number in physics.

D. ABOLITION OF ABSOLUTE SPACE AND TIME

A basic idea of the theory of relativity is the lack of absolute motion, absolute rest, or absolute time. This idea, which Einstein proposed as a scientist, not only violated common sense but also represented a fundamental change in thinking from a philosophical and theological point of view. Among other profound minds of the past, Aristotle, Isaac Newton, and Immanuel Kant all discussed theories of time and space at odds with relativity.

1. Aristotle's View of Space and Motion

The notion of a state of absolute rest was deeply ingrained in Western thought. Aristotle believed that the cosmos was composed of five elements: earth, air, water, fire, and ether. Each of these elements had its "natural place," associated with its divine purpose, and the natural place of all earthlike particles was at the center of the universe. That center was motionless. Indeed, it provided the reference point of absolute rest against which all other motions could be measured. In "On the Heavens," Aristotle writes

God and nature create nothing that does not fulfill a purpose. (p. 31) . . .
All bodies both rest and move naturally and by constraint. A body
moves naturally to that place where it rests without constraint, and rests
without constraint in that place to which it naturally moves. (p. 71) . . .
The natural motion of the earth as a whole, like that of its parts, is
towards the center of the Universe: that is the reason why it is now lying
at the center . . . As evidence that they [heavy objects] move also
towards the center of the earth, we see that weights moving toward the
earth do not move in parallel lines but always at the same angles to it:
therefore they are moving towards the same center, namely that of the
earth. It is now clear that the earth must be at the center and immobile.
. . . From these considerations it is clear that the earth does not move,
neither does it lie anywhere but at the center. (pp. 243–245) [Aristotle,
"On the Heavens," translated by W. K. C. Guthrie, Loeb Classical Library
(Harvard University Press: Cambridge, Massachusetts, 1971)]

2. Newton's View of Space and Time

It is an irony that Isaac Newton believed in the existence of absolute time and
motion even though his own laws of mechanics embodied a principle of rela
tivity. In the beginning sections of his great masterwork, the *Principia* (1687),
Newton writes

Absolute, true, and mathematical time, of itself and from its own nature,
flows equably [equally] without relation to anything external. . . .
Absolute space, in its own nature, without relation to anything external,
remains always similar and immovable. (p. 6)

Lest we think Newton is talking completely in the abstract, he later tells us
that absolute rest and motion are actually distinguishable from relative
motions:

It may be that there is no body really at rest, to which the places and
motions of others may be referred. But we may distinguish rest and
motion, absolute and relative, one from the other, by their properties,
causes, and effects. It is a property of rest that bodies really at rest do rest
in respect to one another. And therefore it is possible that in the remote
regions of the fixed stars, or perhaps far beyond them, there may be
some body absolutely at rest. (pp. 8–9)

For Newton, what defined a state of absolute rest if his own laws of
mechanics did not, if a ball bouncing inside a cart in constant motion behaved
identically to a ball bouncing on the ground? Newton was highly religious,
like most intellectuals of his day, and he equated space to the body of God. It
was God, then, who defined absolute space and absolute rest. As Newton
wrote near the end of the *Principia*

The Supreme God is a Being eternal, infinite, absolutely perfect. . . . He endures forever and is everywhere present; and by existing always and everywhere, he constitutes duration and space. (pp. 544–545) [Isaac Newton, *Principia* (1687), translated by A. Motte and revised by F. Cajori (University of California Press: Berkeley, 1962)]

3. Kant's View of Space and Time

The seminal German philosopher Immanuel Kant (1724–1804) was the son of devout Lutheran parents, who saw to it he received a religious education. As a youngster, studying the Latin classics, Kant became enamored of Lucretius, whom we met in Chapter I, and later studied physics and mathematics as well as theology. Kant was well acquainted with the work of Newton. As he moved toward philosophy, Kant became concerned with the theory of knowledge and how we know what we know. In his landmark philosophical treatise, *The Critique of Pure Reason* (1781), Kant argues that there are fundamental things we understand about the world *prior* to our experience with the world. Indeed, Kant holds that we can relate to the world outside of ourselves only because certain knowledge and concepts are already built into our brains. Falling into this category are concepts of time and space, which lie at the root of all knowledge. Kant begins his *Critique* by asking whether time and space exist independently of our minds:

> What then are time and space? Are they real existences? Or are they merely relations or determinations of things, such, however, as would belong to these things in themselves, though they should never become objects of intuition; or are they [time and space] such as belong only to the form of intuition and consequently to the subjective constitution of the mind, without which these predicates of time and space could not be attached to any object? (p. 24)

By "intuition," Kant means seeing, hearing, feeling, and interaction with the world, and by "subjective constitution of the mind," he means the way in which the mind is organized and thinks. Kant soon answers his questions by deciding that space does not depend on objects external to our minds or on our experience with those objects. Rather, space is a construction and creation of the human mind. That construction, prior to any experience with the world, is essential in order to prepare the mind to conceive of and make sense of objects in the outside world.

> Space then is a necessary representation *a priori* [in advance], which serves for the foundation of all external intuitions [knowledge of the outside world]. . . . Space does not represent any property of objects as things in themselves, nor does it represent them in their relations to each

other. . . . Space is nothing else than the form of all phenomena of the external sense, that is, the subjective condition of the sensibility, under which external intuition is possible. Now, because the receptivity or capacity of the subject [person] to be affected by objects necessarily precedes all intuition of these objects, it is easily understood how the form of all phenomena can be given in the mind previous to all actual perceptions. . . . It is therefore from the human point of view only that we can speak of space, extended objects, etc. (pp. 24–25)

Kant's view of time was similar to his view of space:

Time is not an empirical [observational] conception. For neither coexistence nor succession [in time] would be perceived by us, if the representation of time did not exist as a foundation *a priori*, . . . Time is nothing else than the form of the internal [mental] sense, that is, of the intuitions of self and of our internal state. For time cannot be any determination of outward phenomena. (pp. 26–27) [Immanuel Kant, *Critique of Pure Reason* (1781), translated by J. M. D. Meiklejohn in the Encyclopedia Britannica's *Great Books of the Western World* (University of Chicago Press: Chicago, 1987)]

Discussion Questions III–2

Kant says that time and space are constructions of the human mind. Does the theory of relativity allow for the existence of time outside our minds? Is there a contradiction between the relativity of time—the fact that time flows at different rates for different observers—and the independent existence of time outside of our perception of it?

Discussion Questions III–3

Can you remember your earliest ideas about time and space? Did you have the idea of an absolute condition of rest? If so, where did you get that idea? How do you feel about a condition of absolute rest? Does that idea bring you security, or anxiety, or what?

Discussion Questions III–4

Our intuition and common sense tell us that time flows uniformly, at the same rate for all clocks. On what is this belief based? Have you ever tested this belief? To what accuracy could you prove that time flows at the same rate for all clocks?

Discussion Questions III–5

Most people, including scientists, were not willing to accept the theory of relativity for a long time. As William Magie, Professor of Physics at Princeton, said in a 1911 address to the American Association for the Advancement of Science, "I do not believe that there is any man now living who can assert with truth that he can conceive of time which is a function of velocity or is willing to go to the stake for the conviction that his 'now' is another man's 'future' or still another man's 'past.' " Why do you think people are upset to learn that time does not flow at an absolute rate?

E. EINSTEIN'S APPROACH TO SCIENCE

1. Scientific Postulates as "Free Inventions" of the Human Mind

Einstein had a unique philosophy toward science, extending to his conception of time and space. In Einstein's view, while scientific theories certainly had to be discarded if they were found to disagree with experiment, those theories—and indeed scientific truth—could not be arrived at merely by observing nature. Rather, truths of nature were discovered by inspired guesses, originating in the human mind. In Einstein's words,

> We now know that science cannot grow out of empiricism [experiment and observation] alone, that in the constructions of science we need to use free invention which only *a posteriori* [afterward] can be confronted with experience as to its usefulness. This fact could elude earlier generations [of scientists] to whom theoretical creation seemed to grow inductively out of empiricism without the creative influence of a free construction of concepts. [Albert Einstein in *Emanuel Libman Anniversary Volumes*, vol. 1, p. 363 (International: New York, 1932)]

In his autobiography, Einstein further says that "the system of concepts [about nature] is a creation of man . . ." ["Autobiographical Notes," op. cit., p. 13]. There are subtle but crucial distinctions between Einstein and Kant in their views of the origin of knowledge. As you recall, Kant also believed that certain concepts, such as ideas about time and space, came from the human mind. Einstein, who began reading widely in philosophy as a teenager, had this to say about Kant's views:

> Kant, thoroughly convinced of the indispensability of certain concepts, took them—just as they are selected—to be the necessary premises of every kind of thinking and differentiated them from concepts of empirical origin. I am convinced, however, that this differentiation is erro-

neous, i.e., that it does not do justice to the problem in a natural way. All concepts, even those which are closest to experience, are, from the point of view of logic, freely chosen conventions. ["Autobiographical Notes," op. cit., p. 13]

In this passage, Einstein is saying that there are no decrees about how things must be. Nature is not required to be the way we think it must be, and our minds are not required to think in only certain ways. According to Einstein, until our ideas are framed in a self-consistent theory and tested against experiment, we should consider all possibilities.

Discussion Questions III–6

Kant says that space and time are "the subjective condition of the sensibility" that "necessarily precedes all intuition," and Einstein says that constructions of concepts in science need to use "free invention" of the mind. It sounds as if both men are saying that scientific concepts originate in the human mind. What's the difference in the two viewpoints? Einstein faults Kant for believing that certain concepts such as the nature of time and space are built into our minds, prior to experiencing the world, in a specific and "indispensable" form and that such specific ideas of time and space are "necessary" for how we can think and understand the world. Do Einstein's "free inventions" of the mind require any particular ideas of time and space? Did Einstein believe that specific properties of time and space were givens? For the moment, put yourself in Einstein's shoes. What would you do if experimental tests showed that your two postulates of relativity were false? Would you declare that the experimental results are not relevant to the meaning of time and space, or would you go back to the drawing board and "freely invent" another set of postulates?

In his "Autobiographical Notes," written when he was 67, Einstein described his thinking about the meaning of time and space:

One had to understand clearly what the spatial co-ordinates and the temporal duration of events meant to physics. The physical interpretation of the spatial co-ordinates presupposed a fixed body of reference which moreover had to be in a more or less definite state of motion . . . With such an interpretation of the spatial co-ordinates the question of the validity of Euclidean geometry [and other quantitative notions about space] becomes a problem of physics. If, then, one tries to interpret the time of an event analogously, one needs a means for the measurement of the difference in time. . . . A clock at rest relative to [a system of reference] defines a local time. . . . One sees that *a priori* it is not at all necessary that the "times" thus defined in different [moving] systems agree with one another. ["Autobiographical Notes," op. cit., p. 55]

Discussion Questions III–7

What can you infer about Einstein's attitude toward concepts of time and space from the above passage? What does he mean when he says that one must understand what time and space mean to *physics*? Do you think Einstein had preconceived ideas about the properties of time and space? Could time and space have a meaning outside of physics? Do you think Kant was considering time and space as they related to physics? Do the rulers and clocks that measure time and space intervals have a reality outside of the human mind? How does the above passage help distinguish between Einstein's view and Kant's view of the meaning of time and space?

Discussion Questions III–8

If scientific truths originate from inspired guesses, as Einstein holds, then how can science be an objective activity, in accord with external reality? Discuss the process by which inspired guesses and experiments might work together, back and forth, to arrive at scientific truth.

---

2. Deductive versus Inductive Thinking and the Influence of Hume on Einstein

Associated with Einstein's faith in inspired guesses was his deductive approach to science, an approach differing from that of most other scientists both then and now. Reasoning in science, as in other endeavors, comes in two forms: inductive and deductive. In inductive science, the scientist makes a number of observations of nature, finds a pattern, generalizes the pattern in a "law" or organizing principle, and tests that law against future experiments. For example, the German astronomer Johannes Kepler pored over the data on planets, analyzing the numbers in all kinds of ways, before discovering in 1619 a striking relationship between a planet's distance from the sun and the time it takes to complete an orbit: the square of the orbital period divided by the cube of the distance is the same for all planets. Kepler's law has since been tested and used over and over to predict the orbits of stars and galaxies far beyond the solar system. Darwin traveled to Patagonia, Tierra del Fuega, the Amazon, and Malay, and he spent years examining the vital statistics of coral polyps, ostriches, butterflies, and armadillos before formulating the principle of natural selection.

In deductive science, the scientist begins by postulating certain truths of nature, with only secondary guidance from outside experiments, and deduces the consequences of the postulates. The consequences are cast into predictions, which can then be pitted against observational tests. Inductive reasoning works from the bottom up; deductive from the top down.

In the late nineteenth and early twentieth century, most of Einstein's contemporaries attempted to explain the various observed electromagnetic and mechanical phenomena inductively. As we mentioned earlier, the Dutch physicist Lorentz, reasoning from the results of the Michelson-Morley experiment and other similar experiments, proposed a detailed theory for how the ether would interact with the electrons of bodies to alter their properties. Einstein, however, began with two sweeping postulates far simpler than the detailed theories of Lorentz and deduced their consequences. What Einstein ended up with was a new understanding of time and space.

Of course, all scientific theories of nature, whether they be arrived at by induction or deduction, must be abandoned if they are found to be wrong. However, says Einstein, we will uncover the deep truths not by looking outside and slowly building a dossier of facts, but rather by looking inside our own minds. Although this deductive approach is rare, other twentieth-century physicists have used it successfully. The modern theory of electrons, framed by Paul Dirac in the 1920s, was founded upon Dirac's intuition and love of mathematical beauty, not on observations of how electrons behaved. Likewise, the unified theory of the electromagnetic force and the weak nuclear force, formulated by Sheldon Glashow, Abdus Salam, and Steven Weinberg in the 1960s, was built upon a sense of the unity of nature, not upon detailed trajectories of particles in atom smashers.

For his deductive approach to science, Einstein explicitly credits the Scottish philosopher David Hume (1711–1776). In his autobiography, Einstein writes,

> A remark to the historical development. Hume saw clearly that certain concepts, as for example that of causality, cannot be deduced from the material of experience by logical methods. (p. 13) . . . From the very beginning it appeared to me intuitively clear that, judged from the standpoint of [an observer pursuing a light beam at the speed of light], everything would have to happen according to the same laws as for an observer who, relative to the earth, was at rest. . . . One sees that in this paradox the germ of the special relativity theory is already contained. Today everyone knows, of course, that all attempts to clarify this paradox satisfactorily were condemned to failure as long as the axiom of the absolute character of time, viz., simultaneity, unrecognizably was anchored in the unconscious. Clearly to recognize this axiom and its arbitrary character really implies already the solution of the problem. The type of critical reasoning which was required for the discovery of

this central point was decisively furthered, in my case, especially by the reading of David Hume's and Ernst Mach's philosophical writings. (p. 53) [Albert Einstein, "Autobiographical Notes," op. cit.]

Hume argued that all knowledge of the world based on experience is much less certain than we believe it is. When we witness a moving billiard ball strike another ball at rest and then see the second ball begin moving, we assume a causal relationship between the two events; we assume that the second ball began moving *necessarily* because it was struck by the first. Hume warns that this path to knowledge is dangerous. There is nothing inherent in the motions of the two balls that requires them to be causally connected. It is only because we observe such events frequently that we assume they are causally connected. (To elaborate on Hume's position, one could imagine a hidden device underneath the pool table that stopped the first billiard ball just at the moment it struck the second and another hidden device that happened to kick the second ball at the same moment. In such a situation, the movement of the second ball would indeed not have been caused by the first ball, despite appearances to the contrary.)

As Hume writes in *An Enquiry Concerning the Human Understanding* (1748)

It may, therefore, be a subject worthy of curiosity, to enquire what is the nature of that evidence which assures us of any real existence and matter of fact beyond the present testimony of our senses (20.) . . . All reasonings concerning matter of fact seem to be founded on the relation of *Cause and Effect*. (22.) . . . Suppose a person, though endowed with the strongest faculties of reason and reflection, to be brought [all of] a sudden into this world; he would, indeed, immediately observe a continual succession of objects, and one event following another; but he would not be able to discover anything farther. He would not, at first, by any reasoning, be able to reach the idea of cause and effect; since the particular powers, by which all natural operations are performed, never appear to the senses; nor is it reasonable to conclude, merely because one event, in one instance, precedes another that therefore the one is the cause, the other the effect. (35.) . . . All inferences from experience, therefore, are effects of custom, not of reasoning. (36.) [David Hume, "An Enquiry into Human Understanding" (1748), in Encyclopedia Britannica's *Great Books of the Western World* (University of Chicago Press: Chicago, 1987); numbers in parentheses refer to sections of the treatise, not page numbers]

Discussion Questions III–9

Can you tell why Einstein was influenced by his reading of Hume? Are our commonsense ideas about time and space, based on experience, logically necessary?

Discussion Questions III–10

From the above passage, do you think that Hume believed in the reality of cause and effect relationships? If so, how might those relationships be discovered?

F. THE INFLUENCE OF THE THEORY OF RELATIVITY ON LITERATURE

The idea that time flows at a different rate for different observers is unquestionably profound. Yet one could imagine that many profound ideas might be discussed only among physicists. Not so with relativity. Relativity, like the other ideas in this book, has seeped into our culture and affected us far beyond science. As an illustration here, we will briefly consider the influence of relativity on two major writers of the twentieth century: Vladimir Nabokov and Jorge Luis Borges. Rather than listen to these writers hold forth on relativity, we will see how their consciousness of relativity enters their art.

Vladimir Nabokov (1899–1977), born into an old aristocratic family in Petersburg, Russia, wrote in both Russian and English. His best known novel is *Lolita* (1955). Here, we will excerpt from the novel *Ada* (1969), a lengthy family chronicle. One of the principal characters of the book is Van Veen, a sensuous and eccentric writer. Veen is obsessed with time, among other things. In a long passage, running on for 28 pages, he describes his ideas about time. Some excerpts follow:

> I delight sensually in Time, in its stuff and spread, in the fall of its folds, in the very impalpability of its grayish gauze, in the coolness of its continuum. . . . Time is rhythm: the insect rhythm of a warm humid night, brain ripple, breathing, the drum in my temple—these are our faithful timekeepers; and reason corrects the feverish beat. (pp. 537–538) . . .
> Pure Time, Perceptual Time, Tangible Time, Time free of content, context, and running commentary—this is *my* time and theme. All the rest is numerical symbol or some aspect of Space. (p. 539)

> At this point, I suspect, I should say something about my attitude to "Relativity." It is not sympathetic. What many cosmogonists tend to accept as an objective truth is really the flaw inherent in mathematics which parades as truth. The body of the astonished person moving in Space is shortened in the direction of motion and shrinks catastrophically as the velocity nears the speed beyond which, by the fiat of a fishy formula, no speed can be. That is his bad luck, not mine—but I sweep away the business of his clock's slowing down. Time, which requires the utmost purity of consciousness to be properly apprehended, is the most

rational element of life, and my reason feels insulted by those flights of Technology Fiction. (p. 543)

Perceived events can be regarded as simultaneous when they belong to the same span of attention. . . . I know relativists, hampered by their "light signals" and "traveling clocks," try to demolish the idea of simultaneity on a cosmic scale, but let us imagine a gigantic hand with its thumb on one star and its minimus on another—will it not be touching both at the same time—or are tactile coincidences even more misleading than visual ones? I think I had better back out of this passage. (pp. 543–544) [Vladimir Nabokov, *Ada* (Vintage Books: New York, 1990)]

Discussion Questions III–11

Why does Veen dislike the ideas of relativity? Do you sympathize with his attitude? Do you think Veen dislikes science in general as well as relativity? How do Veen's references to the theory of relativity make him a more interesting character?

Discussion Question III–12

In what ways does Nabokov show a familiarity with the theory of relativity?

Jorge Luis Borges (1899–1986) was an Argentine poet, essayist, short story writer, and literary critic. In 1938, after a head wound and blood poisoning that left him near death, Borges embarked upon his most creative and imaginative stories. We will quote here from the story "The Garden of Forking Paths" (1941), which refers to relativity much more obliquely than *Ada* does. In this story, a Chinese man, Dr. Yu Tsun, is spying for Germany during World War I. Tsun, in England, has decided to kill a well-known Sinologist named Stephen Albert in order to signal Tsun's chief in Berlin that Germany should bomb a British city named Albert. Tsun is at Albert's house, moments before carrying out his deed, when he learns that Albert has in his possession the newly translated manuscript of a book written by Tsun's illustrious great grandfather, Ts'ui Pen. The book is titled *The Garden of Forking Paths*. Unaware of Tsun's intentions to murder him, Albert discusses the book and its author with Tsun:

Philosophical conjectures take up the greater part of his novel. I know that of all problems, none disquieted him more, and none concerned him more than the profound one of time. . . . *The Garden of Forking Paths* is an enormous guessing game, or parable, in which the subject is time. The rules of the game forbid the use of the word itself. . . . *The Garden of Forking Paths* is a picture, incomplete yet not false, of the universe such as Ts'ui Pen conceived it to be. Differing from Newton and Schopenhauer, your ancestor did not think of time as absolute and uniform. He believed in an infinite series of times, in a dizzily growing, ever spreading network of diverging, converging and parallel times. This web of time—the strands of which approach one another, bifurcate, intersect or ignore each other through the centuries—embraces *every* possibility. We do not exist in most of them. In some you exist and not I, while in others I do, and you do not, and in yet others both of us exist. In this one, in which chance has favored me, you have come to my gate. In another, you, crossing the garden, have found me dead. In yet another, I say these very same words, but am an error, a phantom. [Jorges Luis Borges, "The Garden of Forking Paths," translated from the Spanish by Helen Temple and Ruthven Todd, in *Ficciones* (Grove Press: New York, 1962), pp. 99–100]

Discussion Questions III–13

Why would the nature of time be relevant to Tsun, who is about to murder Albert? Discuss the irony in that only moments before his death, Albert talks about being dead in one world and alive in another.

Discussion Questions III–14

If you were a fiction writer, would you be interested in new ideas about the flow of time? If so, why?

G. RELATIVITY AND SCULPTURE

Sculptors and architects have always been especially sensitive to questions of space and spatial relationships. In recent years, sculptors have also taken more interest in *time* as an element of their art. In particular, a new kind of art, called "earth works" or "site sculpture," involves sculpture on a large physical scale, extending over hundreds of meters or even kilometers. In such art, an extended period of time is required for the spectator to walk through and observe the sculpture. Thus, the sculpture unfolds over time. The passage of time becomes part of experiencing the sculpture.

It would be difficult to find clear evidence of the awareness of relativity theory in the construction of earth works and site sculpture, even if such sculpture were standing in front of us. However, it seems likely that relativity has indeed influenced the work of at least some modern sculptors. The following statements by Robert Morris, Richard Serra, and Athena Tacha suggest such an influence.

In an article titled "The Present Tense of Space," first appearing in *Art in America*, sculptor Robert Morris (1931–) wrote

> The '70s have produced a lot of work in which space is strongly emphasized in one way or another. . . . Now images, the past tense of reality, begin to give way to duration, the present tense of immediate spatial experience. Time is in this newer work in a way it never was in past sculpture. Modernist issues of innovation and stylistic radicalism seem to have nothing to do with these moves. More at issue perhaps is a shift in valuation of experience. [Robert Morris, "The Present Tense of Space," in *Art in America*, (January–February 1978), p. 70.]

Athena Tacha (1936–) creates scupltures that one experiences in a rhythmic way. Her "Ripples" (1979), for example, is a series of twisting steps, up and down, measuring 10 meters wide and 27 meters long, built of white concrete. She writes,

> Time enters my sculpture at many levels. One, and perhaps the only way to perceive time is through displacement in *space* of one body in relation to another, i.e., through motion. (p. 216) . . . What led me to develop this range of forming devices was the need to express some of my vital interests. The fundamental importance of gravity in the structure of the universe and in the very nature of space and matter [a reference to Einstein's theory of general relativity]; the interchangeability of matter and energy; the equivalence between acceleration and gravitational pull [another reference to the general theory]; the interdependence of space and time, and the peculiar characteristics of the latter . . . these and other such concepts that modern science has developed are subjects of great excitement to me, which I want to render tangible and communicable to others through the language of form. (p. 218) [Athena Tacha, in *American Artists on Art*, ed. E. H. Johnson (Harper & Row: New York, 1982); reprinted from Athena Tacha, "Rhythm as Form" *Landscape Architecture* (May 1978)]

The site sculptor Richard Serra (1939–) has explained how his sculpture is formed by the site and the experience of walking through the site. His piece "Shift" (1970–1972) is located in a farming field in King City, Canada. The sculpture consists of six cement sections, each 1.7 meters thick. The six sections are placed in a zigzag manner and span a distance of 250 meters. Serra describes his work:

> The boundaries of the work became the maximum distance two people could occupy and still keep each other in view. . . . What I wanted was

a dialectic between one's perception of the place in totality and one's relation to the field as walked. . . . The intent of the work is an awareness of physicality in time, space, and motion. Standing at the top of the eastern hill, one sees the first three elements in a Z-like linear configuration. The curvature of the land is only partially revealed from this point of view, because the configuration compresses the space. Until one walks into the space of the piece, one cannot see over the rise, as the hill descends into its second and third 5-foot drop. (p. 208) [Richard Serra in *American Artists on Art*, op. cit.; reprinted from Richard Serra, "Shift" in *Arts Magazine* (April 1973)]

Discussion Question III–15

How do you think the notions of time and space as expressed by relativity might affect the thinking of an artist?

Discussion Questions III–16

From the passages quoted, do you think that Nabokov and Tacha have an understanding of the theory of relativity? To what extent might a writer or an artist need to understand the ideas of relativity in order to be affected by it? What concepts could be understood without a detailed and quantitative study of the subject?

Discussion Questions III–17

Seonaidh Davenport, who graduated from Princeton in 1990 as an art major, argues that while art is undoubtedly affected by science, an artist too concerned with the details of science will not produce good art. Do you agree? Discuss your views on the positive and negative ways that ideas in science might influence artists and art.

Readings

Jorges Luis Borges, "The Garden of Forking Paths," trans. from the Spanish by Helen Temple and Ruthven Todd, in *Ficciones* (Grove Press: New York, 1962).

Albert Einstein, "Physics and Reality" (1936), in Albert Einstein, *Ideas and Opinions* (Dell: New York, 1981).

Albert Einstein, "Autobiographical Notes," in *Albert Einstein: Philosopher-Scientist*, ed. P. A. Schilpp (The Library of Living Philosophers: Evanston, Illinois, 1949).

Albert Einstein, "Electrodynamics of Moving Bodies" (1905) in H. A. Lorentz, A. Einstein, H. Minkowski, and H. Weyl, *The Principle of Relativity* (Dover: New York, 1953) [This is Einstein's landmark paper on the theory of special relativity, translated into English. The paper is technical. However, the first few pages are very readable.]

David Hume, "An Enquiry into Human Understanding" (1748), in Encylopedia Britannica's *Great Books of the Western World* (University of Chicago Press: Chicago, 1987).

Immanuel Kant, *Critique of Pure Reason* (1781), translated by J. M. D. Meiklejohn in the Encyclopedia Britannica's *Great Books of the Western World* (University of Chicago Press: Chicago, 1987).

Robert Morris, Richard Serra, and Athena Tacha in *American Artists on Art*, ed. E. H. Johnson (Harper & Row: New York, 1982).

Vladimir Nabokov, *Ada* (McGraw-Hill: New York, 1969).

WERNER HEISENBERG

Werner Karl Heisenberg (1901–1976) was born in Würzburg, Germany and received his Ph.D. in theoretical physics in 1923 at the University of Munich. Beginning in 1924, he spent three years working with the pioneering atomic physicist Niels Bohr in Copenhagen.

In the middle and late 1920s, Heisenberg developed a mathematical theory to account for the newly observed wave-particle duality of nature. At the same time, the physicist Erwin Schrödinger developed a different theory to describe the same phenomena. Later, both theories—which together came to be called quantum mechanics—were shown to be equivalent, in the sense that they made identical predictions. Yet the two theories were quite different in their mathematics and intuitive associations. Heisenberg's theory was much more abstract. Indeed, Heisenberg decided that quantum phenomena demanded such a radically new view of nature that any attempt to visualize the subatomic world was doomed to failure and would lead to erroneous conceptions. In a paper in 1926, he wrote that the new theory of quantum mechanics worked "under the disadvantage that there can be no directly intuitive geometrical interpretation because the motion of electrons cannot be described in terms of the familiar concept of space and time." In another paper of the same year, he wrote that the "electron and the atom possess not any degree of direct physical reality as the objects of daily experience." Yet, at the same time, Heisenberg said that "any experiment in physics, whether it refers to phenomena of daily life or to atomic events, is to be described in the terms of classical [pre-quantum] physics." Heisenberg recognized the necessity, but inadequacy, of familiar concepts and language.

In 1937, Heisenberg married Elisabeth Schumacher and with her had 7 children. He was deeply interested in music and philosophy as well as physics. In World War II, Heisenberg remained in Germany, unlike some of his German colleagues, and contributed to the German effort to develop nuclear weapons. For this work, he was later held in contempt by other scientists and was sometimes booed and hissed at when he gave lectures outside Germany.

Heisenberg won the Nobel prize for his work in quantum mechanics in 1932, at the age of 31, making him among the youngest physicists to win the prize. (The youngest was W. L. Bragg, who won the prize at age 25; next youngest were Heisenberg, Carl Anderson, P. A. M. Dirac, and T. D. Lee, all of whom won the prize at age 31. The average age is in the late forties.)

CHAPTER 4

The Wave-Particle Duality of Nature

Indeed, we find ourselves here on the very path taken by Einstein of adapting our modes of perception borrowed from the sensations to the gradually deepening knowledge of the laws of Nature. The hindrances met on this path originate above all in the fact that, so to say, every word in the language refers to our ordinary perception.

Niels Bohr, *Nature (Supplement)*, 14 April 1928, p. 580

Soon after relativity came the second great revolution in twentieth-century physics: quantum physics, or the discovery of the wave-particle duality of nature. Like the relativity of time, the wave-particle duality of nature violates common sense. According to this duality, a piece of matter—an electron, for example—behaves sometimes as if it were in only one place at a time, like a particle, and sometimes as if it were in several places at the same time, like a wave on a pond. Indeed, our understanding of objects based on our ordinary perceptions of the world are no longer valid.

The wave-particle duality of nature has strange consequences. For one, whether an object behaves like a particle or like a wave depends on how we choose to observe it. In other words, our act of observation seems to determine the properties of the object. Before we observe an object, it is impossible to say whether the object is a wave or particle or even if the object exists at all. Once we observe an object, we cannot disentangle ourselves from the object. The observer and observed are tied together in an inseparable knot. Furthermore, the wave-particle duality of nature leads to an essential indeterminism in science, an essential inability to predict precisely the future state of affairs of a system. Both of these new outcomes—the inseparability of the observer and the observed, and the essential indeterminism of nature—violate not only common sense but also the most fundamental notions of prequantum science.

Quantum physics was discovered and developed in the first three decades of this century. The theory has been brilliantly confirmed in the laboratory. Yet physicists are still baffled by quantum physics, even though they have no trouble calculating its consequences. The phenomena of relativity also contradict ordinary intuition, but those contradictions seem tame by comparison with the dilemmas of quantum physics. What is matter? How can an object be at many places at the same time? Does matter have an existence independent of our observation of it?

A. WAVES

Before we begin our discussion of the wave-particle duality of nature, we will first have to be discuss the concept of a particle and that of a wave. A particle is easier to define, since it corresponds closely to our idea of "object," like a billiard ball or a raindrop. At any moment in time, a particle occupies a single, localized region of space and has its energy concentrated in that region. A particle has boundaries. With a particle, you can say where it is and where it isn't. As we will now see, we cannot make these statements about waves.

1. Definition of a Wave

Roughly speaking, a wave is a pattern of matter or energy that is spread over a volume of space. Examples of waves are ripples on a pond (water waves), vibrating violin strings, and oscillating masses of air (sound waves). By contrast to particles, waves cannot be confined to localized regions of space, but instead have their mass and energy spread out over a sizable volume. In a water wave, for example, each successive trough lies in one place and each crest in another. The following sections will define waves more precisely.

2. Properties of Waves: Wavelength, Frequency, Speed, and Amplitude

In this section and for most of the chapter we will consider simple waves called monochromatic waves. Such waves have a repeating pattern. A snapshot of a monochromatic wave shows a pattern such as in Fig. IV–1a. If Fig. IV–1a represents a water wave, for example, then the wavy line gives the height of the water at a moment in time. The distance between two successive crests is called the *wavelength* of the wave, often denoted by the symbol λ. Since the pattern of a monochromatic wave is symmetrical about each crest and trough, the distance between two successive troughs, or between any other two successive corresponding points, is the same. Now we can better define a monochromatic wave: a wave that contains only a single wavelength, with exactly the same distance between any two crests or any two troughs. When waves have a mixture of wavelengths, they have more complicated patterns.

Wavelengths of water waves range from fractions of a centimeter up to kilometers. Electromagnetic waves, which we considered in Chapter III, come

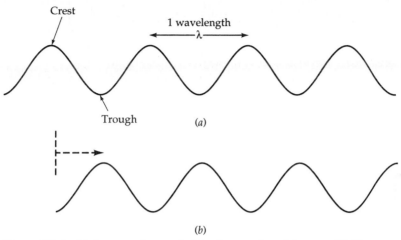

Figure IV–1: (*a*) A monochromatic (single-wavelength) wave. (*b*) The same wave a moment later, after it has traveled to the right.

in an enormous range of wavelengths, only a small fraction of which interact with the human eye to produce the sensation of visible light. Each different color of light corresponds to a different wavelength. (This is the origin of the term "monochromatic" for waves containing only a single wavelength.) Yellow light, for example, has a wavelength of about 0.000058 centimeters. Wavelengths of light smaller than about 0.000039 centimeters or larger than about 0.000077 centimeters cannot be seen with the eye.

For a moving wave, such as a traveling water wave, the crests and troughs move in time. If you could travel alongside the wave, at the same speed as the wave, you would see a fixed pattern, as if you were running along side a moving train and stayed even with the dining car. If you remained at one position in space, however, the pattern of the wave would change, with a crest moving by, then a trough, then another crest, just as when a train goes by and different cars pass you one by one. Figure IV–1b shows the wave in Fig. IV–1a at a later moment of time. Note that the pattern of the wave has shifted to the right; the wave has traveled to the right. The dashed vertical line in Fig. IV–1b shows where the first crest of the wave was at the earlier moment, when the snapshot in Fig. IV–1a was taken. The dashed horizontal arrow shows how far the wave has traveled.

The number of crests that pass a fixed point in space each second is called the *frequency* of the wave. Frequency is measured in cycles (crests) per second and often denoted by the symbol v.

The speed of any particular crest of a moving wave is called the speed of the wave. Since the entire pattern of a traveling wave moves together, all crests and troughs move at the same speed.

Finally, the strength of a wave is called its *peak amplitude*. For a water wave, the peak amplitude is the maximum height of the water above the undisturbed level of the lake or pond. Strong waves are high, with large peak amplitudes, and weak waves are low, with small peak amplitudes.

For a traveling wave, there is a simple relationship between wavelength λ, frequency v, and speed v. Since v is the number of crests that pass a fixed point in space each second, a new crest must pass the point every $1/v$ seconds. For example, if $v = 3$, three crests pass by every second, meaning a new crest passes by every $1/v = 1/3 = 0.33$ seconds. Now, each crest must move a distance λ to get to where the crest just ahead of it used to be, since λ is the distance between crests. This movement happens during a time interval $1/v$, which is the time for two successive crests to pass the same point. Thus the wave crest's speed, which is the distance a crest travels divided by the time it takes to travel that distance, is

$$v = \frac{\lambda}{1/v} = \lambda v. \tag{IV–1}$$

Since the speed of each crest is the speed of the wave, Eq. (IV–1) gives the speed of the wave.

For a light wave, or for any other electromagnetic wave, the speed of the wave is $v = c = 299{,}793$ kilometers per second. In this chapter, will round this number to 300,000 kilometers per second, or 3×10^8 meters per second. For a wave traveling at speed c, Eq. (IV–1) gives a relationship between wavelength and frequency, $c = \lambda v$, or dividing by v,

$$\lambda = \frac{c}{v}. \tag{IV–2}$$

3. Interference of Waves

If two or more waves overlap, they can enhance each other or cancel each other out, depending on how the crests and troughs of the different waves line up with each other. This process is called *interference* and is a key identifying feature of waves. When waves enhance each other, the process is called constructive interference; when they cancel, the process is called destructive interference.

For example, consider two waves of the same wavelength and peak amplitude moving in the same direction. If the two waves line up such that the troughs of one wave are aligned with the crests of the other, as shown in Fig. IV–2a, the two waves will exactly cancel each other. The net result will be no wave at all. If these were water waves, the cancellation of the two waves would leave the water surface flat. In fact, you can sometimes see the cancellation of water waves on a pond, when two sets of ripples overlap and leave a

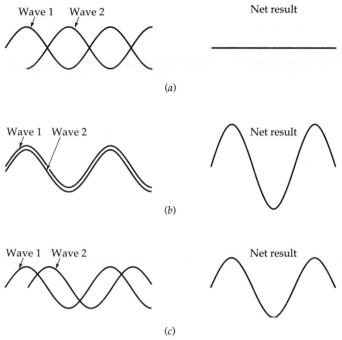

Figure IV–2: (*a*) The destructive interference of two waves completely misaligned with each other. (*b*) The constructive interference of two waves completely aligned with each other. (*c*) The partially constructive interference of two waves partly aligned with each other.

calm area in between. At the other extreme, if the crests of the two waves coincide, the two waves will enhance each other as much as possible. The net result is a wave of the same wavelength but twice the peak amplitude, as shown in Fig. IV–2b. Between these two extremes of complete cancellation and complete enhancement are various intermediate cases, where a crest of one wave lands somewhere between a trough and crest of the other. Such an intermediate case is shown in Fig. IV–2c, where the net result of the two waves is a wave with a peak amplitude of 1.4 times that of either individual wave.

The interference of waves has been observed in many different phenomena, from water waves to sound waves to light waves. Engineers who design ships, concert halls, and precision optical equipment all must take into account the interference of waves.

4. Nonlocality of Waves

It is now easier to understand what we mean when we say that a wave is a "nonlocal" phenomenon. The process of interference requires that waves have crests and troughs, which can cancel or enhance each other. Crests and troughs, in turn, require that waves extend over a region of space of at least several wavelengths in size. It makes no sense to say that a wave exists only at a single position in space; it exists everywhere its crests and troughs are. In this regard, waves differ from particles, whose existence can be limited to single positions in space.

5. Light as a Wave

In Section B–3 of Chapter III we mentioned that in the nineteenth century physicists discovered that light is a traveling wave of electromagnetic energy. In fact, it had been known since the mid-seventeenth century that light behaves like a wave. The wave nature of light was first demonstrated by the Italian Jesuit priest Franceso Maria Grimaldi (1618–1663) and later by the British scientist and linguist Thomas Young (1773–1829). Young showed that light passing through two thin slits in the window shade of a darkened room, upon reaching a screen, produced a curious pattern of light and dark stripes, as shown in Fig. IV–3. If light traveled only in straight lines, then the expected pattern on the screen would be two stripes of light, each illuminated by the light rays passing through the corresponding slit in the window shade. This expected pattern is shown in the leftmost sketch in Fig. IV–3b.

What is actually seen on the screen is shown in the rightmost sketch in Fig. IV–3b. Evidently, the incoming light, upon reaching the two openings in the shade, breaks up into two sets of secondary ripples, just as water rushing past a log in a pond emerges in a series of ripples (Fig. IV–3c). As shown schematically in Fig. IV–3a, the outgoing ripples (waves) of light can overlap and interfere with each other "downstream," in the space between the shade and the screen. At some points on the screen, the overlapping waves reaching those points interfere constructively, and a bright stripe appears; at other points, the waves interfere destructively, and there is no light at all (dark stripes appear). Where there should be only shadow, directly in the middle of the screen and opposite the midpoint between the two slits in the shade, there is an invasion of light; where there should be light, directly behind either slit, there could be darkness. These results follow from the interference of the two outgoing waves of light and show that light is a wave.

B. THE PHOTOELECTRIC EFFECT

1. The Nature of the Electron

In 1897 the British physicist Joseph John Thomson (1856–1940) found that electricity is produced by discrete particles. The particles are called electrons. Each electron has a tiny mass, of about 10^{-30} kilograms, thousands of times smaller

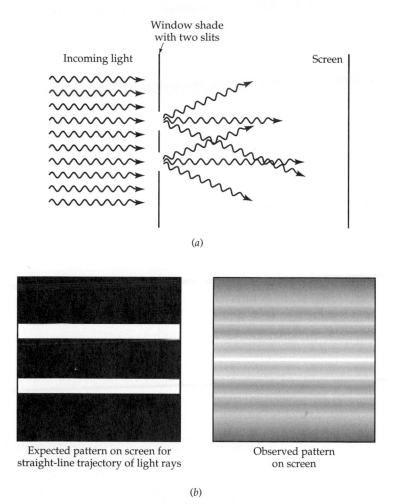

(a)

Expected pattern on screen for
straight-line trajectory of light rays

Observed pattern
on screen

(b)

Figure IV–3: (a) The spreading out of light waves after passing through two thin slits
in a window shade. (b) Patterns of light on a screen after passing through
the slits in part (a); on the left is the pattern expected if the light did not
spread out; this pattern is not observed. On the right is the actually
observed pattern.

than the mass of a typical atom. When electrons move, they produce an electrical current. Indeed, the strength of an electrical current is proportional to
the number of electrons that pass a region of space, such as a wall plug, each
second.

To be more exact, electrons carry a negative electrical charge. Other subatomic particles, such as protons, carry a positive electrical charge. We will be
concerned here only with electrons.

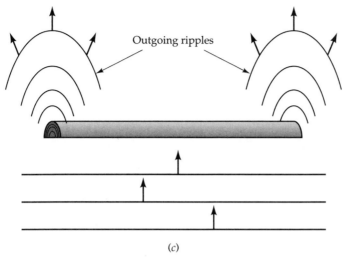

(c)

Figure IV–3: (c) A water wave passing by a log in a pond. Note the outgoing ripples downstream from the log.

2. Discovery of the Photoelectric Effect

It has been known since the late nineteenth century that under some conditions a metal releases electrons when light shines on it, as illustrated in Fig. IV–4. This phenomenon is called the *photoelectric effect*. The number of electrons released per second can be measured by the electrical current produced in a wire connected to the metal. In addition, the kinetic energy of individual electrons can be measured by the force needed to stop them. In terms of energy, the energy of the incoming light beam is converted to the energy of the ejected electrons.

In 1902, the German experimental physicist Philipp Lenard (1862–1947) discovered a number of important properties of the photoelectric effect. First, Lenard found that a greater intensity of incoming light releases more electrons from the metal. This result was expected. If we think of the electrons as particles of sand lodged in a sand bar, then increasing the strength of an incoming wave should dislodge more sand. Secondly, Lenard found that the kinetic energy of individual escaping electrons *does not increase* when the intensity of the incoming light is increased. For light of a given frequency, the kinetic energy of each outgoing electron is *independent* of the intensity of the incoming light. This result was astonishing. Physicists had believed that the kinetic energy of individual electrons should increase with increasing light intensity, just as a stronger wave hitting a sandbar pounds each grain of sand with more force and dislodges it with more speed. Finally, Lenard found that the kinetic energy of individual escaping electrons *does* increase with increasing frequency of the incoming light wave. Quantitatively, he found the result

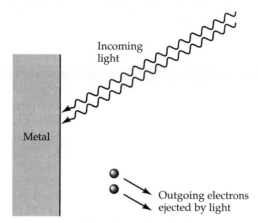

Incoming
light

Metal

Outgoing electrons
ejected by light

Figure IV–4: The photoelectric effect.

$$\tfrac{1}{2}\,mv^2 = h\nu - K, \qquad\qquad\qquad (\text{IV–3})$$

where m and v are the mass and velocity of an escaping electron, $mv^2/2$ is the usual formula for kinetic energy, v is the frequency of the incoming light wave, h is a fixed number, and K is a number that varies from one metal to the next but does not depend on the incoming light. To be more precise, the velocity that occurs in Eq. (IV–3) is the maximum velocity in any large group of emerging electrons for a given metal and frequency of incoming light.

Problem IV–1: Measurement of Photoelectric Constants

Describe how the constants h and K could be determined.

Solution: To determine h, experiment with a single metal. Pick any two frequencies v_1 and v_2 for incoming light. Measure the maximum speed of electrons ejected by incoming light of frequency v_1 and the speed of an electron ejected by incoming light of frequency v_2. Let these two speeds be v_1 and v_2, respectively. According to the experimental result of Eq. (IV–3),

$$\tfrac{1}{2}mv_1^2 = h\nu_1 - K,$$

$$\tfrac{1}{2}mv_2^2 = h\nu_2 - K.$$

If the second equation is subtracted from the first, we get

$$\tfrac{1}{2}m(v_1^2 - v_2^2) = h(v_1 - v_2),$$

which can be solved for h to give

$$h = \frac{m(v_1^2 - v_2^2)}{2(v_1 - v_2)}. \qquad \text{(IV–4)}$$

Notice that all the quantities appearing on the right-hand side of Eq. (IV–4) are *measured* quantities, so that Eq. (IV–4) does indeed determine the constant h. The measured value of h is

$$h = 6.6 \times 10^{-34} \text{ joules} \times \text{seconds}. \qquad \text{(IV–5)}$$

To see that the constant h must have units of joules $\times$ seconds, note from Eq. (IV–4) that h has units of kinetic energy, or joules, divided by frequency. Frequency v is measured in cycles per second, so $1/v$ has units of seconds per cycle. But cycles are just numbers of complete turns, not units like meters or kilograms, so the units of $1/v$ are equivalent to seconds. Hence, h has units of joules multiplied by seconds, or joules $\times$ seconds.

We will be using the fundamental constant h many times in this chapter. Do not confuse it with *height*, which we have denoted by the same symbol h in the past. (In this chapter, we will denote vertical height by the symbol y.) The constant h, given in Eq. (IV–5), plays a key role in quantum theory.

Once h has been determined, K can be determined for any metal by measuring the velocity v of an escaping electron for a known frequency v of incoming light shining on that metal. Then, by Eq. (IV–3),

$$K = hv - \tfrac{1}{2}mv^2.$$

For example, for copper, $K = 6.5 \times 10^{-19}$ joules and for zinc, $K = 4.9 \times 10^{-19}$ joules.

---

3. Einstein's Photon Theory of Light

Some of Lenard's discoveries were most puzzling. If light is a wave, as strongly suggested by interference phenomena, shouldn't a greater intensity of waves impart a greater energy to each ejected electron?

In 1905, the same year that he published his theory of relativity, Einstein proposed a new theory of light, a theory that explained Lenard's results. Einstein proposed that light is not distributed evenly over a region of space, as one would expect from a wave, but instead comes in individual "drops" of energy. Each drop of light is called a photon. Photons act like particles. A typical light beam should be pictured as a flow of raindrops, with empty space between the drops, rather than as a continuous wave of water. Einstein proposed that in a light beam of frequency v, the energy of each photon of light is

$$E = hv. \tag{IV–6}$$

In this new picture of light, an individual electron is hit not by a continuous stream of energy, but by an individual photon of light. Most electrons are not hit by any photons at all and so are unaffected by the incoming light. If a particular electron is struck by and absorbs a photon, it acquires the photon's energy, hv. Some of the energized electrons will bounce around in the metal and completely dissipate their energy, and some may escape the metal, with energy left over. The chance of an energized electron being struck by additional photons, which are relatively few and far between, is extremely small. Einstein interpreted the number K in Lenard's experimental result, Eq. (IV–3), to be the *minimum* energy needed for an electron to break free from the atoms in the metal. Thus an electron struck by a photon gains an energy hv from the photon, uses up at least an energy K to escape the atom, and has a remaining kinetic energy of at most $hv - K$. The fastest-moving escaping electrons should then have a kinetic energy of $hv - K$, as found by Lenard and shown in Eq. (IV–3).

Einstein's photon theory of light explained all of Lenard's results. An increase in light intensity corresponds to a greater number of photons per second. Since each photon may eject an electron, a greater number of electrons will be ejected each second. However, for light of a fixed frequency, the energy of each photon remains the same, $E = hv$, and is *not* increased by increasing the intensity of light. So when a single electron is hit by a single photon, the energy gained by the electron will remain the same, regardless of the intensity of the light. There may be many more photons flying about, but an electron can be hit by only one of them at a time. On the other hand, increasing the *frequency* of the incoming light, whatever its intensity, will increase the energy of each photon of light, as described by Eq. (IV–6), and each ejected electron will gain more energy. Even if the intensity of the incoming light is decreased, so that the light is very dim, an increased frequency gives each photon more energy and thus gives each ejected electron more energy. Perhaps most strikingly, for sufficiently low frequencies of incoming light, no electrons are ejected regardless of the light intensity. This result is, in fact, observed. A high intensity of light means many photons per second, but the energy of each of those photons can still be very small if the frequency of light is low.

Problem IV–2: Minimum Frequencies

What is the minimum frequency of light below which no electrons can be ejected from a metal by the incoming light?

Solution: Since K is the energy needed by an electron to escape a metal and $h\nu$ is the energy gained from an incoming photon, the ejection of an electron from the metal requires that $h\nu$ *be bigger than* K. The minimum frequency for ejection is thus

$$\nu_{min} = K/h.$$

In Einstein's theory of light, light acts like a particle, not like a wave. The constant h, given in Eq. (IV–5), is a fundamental constant of nature, meaning that it has the same value for all situations, in all places and at all times, just as the speed of light is believed to be the same everywhere in the universe. The constant h is called Planck's constant, named after the great German theoretical physicist Max Planck (1858–1947). (Planck, Lenard, and Einstein were all Nobel prize winners.) In 1901, Planck had proposed that an individual atom vibrating at a frequency ν could emit energy not in a continuous range of energies but only in *multiples* of $h\nu$; that is, an atom vibrating at ν cycles per second could emit an energy of $1h\nu$, or $2h\nu$, or $3h\nu$, and so on, but not anything less than $h\nu$ and not any fractional multiple of $h\nu$. Planck had been forced to this odd proposal in an attempt to explain the observed radiation from hot objects like furnaces. From the observations, Planck was able to determine the required value of h for his theory, and it was the same as that later found in the photoelectric effect.

What Planck and Einstein suggested was that energy in nature comes not in a continuous, infinitely divisible stream but in indivisible packets. The indivisible packet of energy is called the *quantum*. The quantum of light is the photon. Analogously, the quantum of money in the United States is the cent. Every purchase involves a multiple of cents, but not fractions of cents.

Problem IV–3: Quanta of Light

Orange light has a frequency of about 5×10^{14} cycles per second. (a) What is the wavelength of orange light? (b) How much energy is carried by one photon of orange light? (c) How many photons of orange light would it take to raise a penny by a distance of 1 meter?

Solution: (a) From Eq. (IV–2),

$$\lambda = \frac{c}{\nu} = \frac{3 \times 10^8 \text{ m/s}}{5 \times 10^{14} / s} = 6 \times 10^{-7} \text{ m} = 6 \times 10^{-5} \text{ centimeters.}$$

(b) From Eq. (IV–6) and Eq. (IV–5),

$$E = h\nu = (6.6 \times 10^{-34} \text{ j} \times \text{s})(5 \times 10^{14}/\text{s}) = 3.3 \times 10^{-19} \text{ joules.}$$

Note that we have used the abbreviation j for joule, just as we use the abbreviation m for meter, kg for kilogram, and s for second.

(c) A penny has a mass of about 0.0025 kilograms. From our expression for gravitational energy in Chapter I, the energy needed to raise a penny by a height of 1 meter is

$$E = mg \, \Delta y = (0.0025 \text{ kg})(9.8 \text{ m/s/s})(1 \text{ m}) = 0.0245 \text{ joules.}$$

Since each orange photon has an energy of 3.3×10^{-19} joules and the energy needed to raise the penny 1 meter is 0.0245 joules, the number of orange photons needed to raise the penny is

$$\frac{0.0245 \text{ j}}{3.3 \times 10^{-19} \text{ j/photon}} = 7.4 \times 10^{16} \text{ photons.}$$

The huge number of photons required for such a menial task shows how small is the energy of each photon. For perspective, a normal sized room, illuminated only by a standard 100 watt light bulb, has about 3×10^{12} photons in it at any one time.

Because the packet of energy, the quantum, is typically so small, we are not aware from ordinary experience that such a minimum amount exists, in the same way that we cannot see the individual grains of sand when looking at a beach from a distance of a few meters or more.

Einstein's photon theory applies to light containing several different frequencies as well as to light of a single frequency. Any composite beam of light can be broken up into its component frequencies—by a prism, for example—and each single-frequency component behaves like a group of photons of that frequency. Thus, a beam of white light, containing many different frequencies, contains photons of many different individual frequencies and individual energies.

Lenard's experimental results with the photoelectric effect and Einstein's interpretations of those results strongly suggested that light came in particles. Yet previous experiments, particularly those involving interference effects, had shown that light behaved like a wave. How could something behave both like a particle, coming in a localized unit, and like a wave, coming in a continuous and spread-out form? This self-contradictory duality dumbfounded many physicists of the day, including Planck and Einstein, and it still does.

4. Verification of the Photon Nature of Light

In 1923, the American physicist Arthur Compton (1892–1962) performed important experiments that helped confirm the photon nature of light. Compton illuminated electrons in carbon and other light elements with X-ray light and measured the angle and frequency of the reflected light. When the bombarding light is of high frequency, as is the case for X-rays, the electrons behave as if they were unconfined by atoms. Instead of absorbing light, as in the photoelectric effect, unconfined electrons scatter light, diverting it to a new direction, as shown schematically in Fig. IV–5a. Compton found that the frequency of the scattered light was always less than the frequency of the incoming light and that the frequency decrease depended in a precise way upon the angle of scattering.

Figure IV–5: (*a*) Scattering of light waves by an electron, known as Compton scattering. (*b*) The same as in part (*a*), with the light represented as photons instead of as waves.

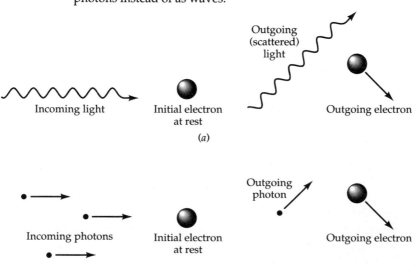

According to the wave theory of light, the scattered light should have had the same frequency as the incoming light. According to the photon theory, however, the results were completely explained. The scattering of light—in microscopic detail—is just like many collisions of billiard balls, two balls at a time. One ball is an electron, initially at rest, and the other is an incoming photon, as shown schematically in Fig. IV–5b. Just as an incoming billiard ball hitting a stationary ball imparts some energy to the latter, thereby losing energy itself, a photon scattering from an electron at rest would be expected to lose energy. The laws of mechanics show how much energy an incoming billiard ball loses for each different angle of rebound. (For example, a 180 degree rebound, in which the incoming ball bounces straight backward, corresponds to the maximum energy lost.) The same laws of mechanics, when applied to the collisions of photons and electrons, exactly agreed with the observed relation between the frequency of scattered light and the angle of scattering. After Compton's work, it was clear that in some experiments light behaved like a group of particles, even though in other experiments it behaved like a wave. Evidently, light has a split personality.

Notice that in Fig. IV–3a, IV–4, and IV–5a, we depicted light as a group of wavy lines, suggesting its wavelike character, while in Fig. IV–5b we depicted light as a group of little balls, suggesting its particle-like character. Since light behaves *both* like a wave and like a particle, it is hard to know how to draw light or how to visualize it. This dilemma is at the core of the wave-particle duality of light, and of nature in general. In the future, we will sometimes draw light as a wave and sometimes as particle, depending on the particular effect being discussed, but we should keep in mind that none of these drawings fully depict the strange character of light.

C. THE DOUBLE-SLIT EXPERIMENT

1. Description of the Experiment

There is a simple but famous experiment that illustrates the wave-particle duality of nature in its most disturbing form. This experiment has several parts. In the first part, put a window shade with a thin horizontal slit in it between a light source and a screen, as shown in Fig. IV–6a. Darken the room, so that the only source of light comes from behind the shade (to the left of the shade, in the figure). Furthermore, make the source of light extremely weak, so that it emits only a few photons of light per second. Now, measure the pattern of light that hits the screen.

Next, cover up the first slit in the shade and cut a second slit above it, as shown in Fig. IV–6b. Repeat the experiment and measure the pattern of light on the screen. In each of these first two experiments, light can get to the screen only through one slit in the shade because only one slit is open at a time.

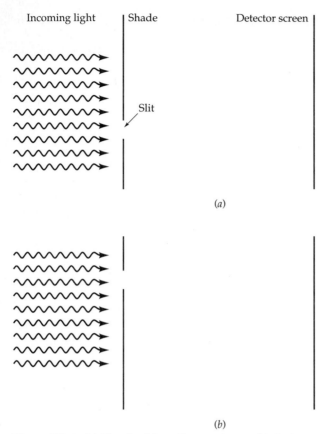

Figure IV–6: (*a*) The double-split experiment. Light is represented as waves. (*b*) The same as part (*a*), with the slit in a different location.

For the third experiment, uncover *both* slits in the shade. If light consists of particles, as the photoelectric effect and Compton's experiments show, then each photon of light coming from behind the shade and striking the screen must pass through *either* the top slit *or* the bottom slit. This seems obvious. A particle can't be in two places at the same time. The photons that pass through the bottom slit should produce a pattern of light on the screen that is identical to that found in the first experiment, where only the bottom slit was open. Likewise, the photons that pass through the top slit should produce a pattern of light on the screen that is identical to that found in the second experiment, where only the top slit was open. Therefore, the pattern of light on the screen in the third experiment should be the *combination* of the patterns seen in the first two experiments. Regions of the screen that were lit up in *either* of the first two experiments should be lit up in the third experiment. Regions of the screen that were dark in both of the first two experiments—that is, regions that were not struck by any photons in either experiment—should be dark in the third experiment.

These anticipated results for the third experiment will indeed be observed if we place photon detectors after the slits in the shade, as shown in Fig. IV–6c, to show us which slit each photon passes through. (Notice that in this figure, we have drawn the incoming light as a group of particles, rather than incoming waves.) Our detectors are designed such that they make a clicking sound each time a photon passes through them. The first such detectors able to detect individual photons were called photomultipliers and were developed in the late 1930s.

If we perform the third experiment with the two photon detectors in place, we can listen for the clicks and know for sure which slit each photon passes through. Remember that our source of light behind the shade is

Figure IV–6: (c) The double-slit experiment, with both slits open and monitored by photon detectors. Here, the incoming light is represented as photons. (d) The double-slit experiment, with both slits open and without the photon detectors. Here, the incoming light is represented as waves.

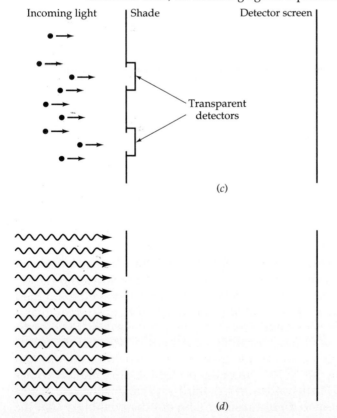

extremely weak and emits only a few photons per second. With the detectors in place, we will sometimes hear one detector click, meaning a photon has just passed through it, and sometimes hear the other detector click, meaning a photon has passed through it. We never hear the two detectors click at the same time. We can literally count photons of light, one at a time, and say for certain which slit each photon passes through on its way to the screen. When we examine the screen, we will indeed find the added patterns of the first two experiments. All of this makes perfect sense and clearly suggests that light is a group of particles.

The trouble comes in the fourth experiment. Keep everything the same, but remove the two detectors, as shown in Fig. IV–6d. Common sense says that the pattern of light on the screen should be the same as in the previous experiment: just because we don't record which slit each photon passes through doesn't mean that the photons don't pass through the slits just as before. With or without the two detectors, the photons still must pass through the slits to illuminate the screen. However, the pattern of light on the screen *does* change. The pattern is no longer the sum of the patterns from each individual slit, found in the first two experiments. Instead, the pattern is now what we would expect from the interference of two overlapping waves emanating from the slits, as illustrated in Fig. IV–3a and the right side of Fig. IV–3b. (Accordingly, we have chosen to draw the incoming light in Fig. IV–6d as incoming waves.) This version of the experiment was first performed about 1807 by Thomas Young, whom we have mentioned earlier.

Let's think about what this means. Interference of two waves requires that the waves overlap, that they occupy the same space. Thus, *two* waves of light must simultaneously occupy the space between the shade and the screen during the period of transit from the shade to the screen. For the sake of calculation, let's say that the shade and the screen are 1 meter apart. Since light travels with a speed of about 3×10^8 meters per second, the passage time from shade to screen is about

$$\frac{1 \text{ meter}}{3 \times 10^8 \text{ m/s}} = 3.3 \times 10^{-9} \text{ seconds.}$$

From this calculation, in order to produce the observed interference pattern on the screen, the two light waves, one from each slit, must be emitted no further apart in time than every 3.3×10^{-9} seconds, so that they will overlap on their way to the screen. However, we have made our source of light so weak that it emits only a few photons per second—let's say one photon every 0.3 seconds. So two successive photons cannot possibly overlap on their way from the shade to the screen. Each photon has long since emerged from the slit and traveled to the screen, in 3.3×10^{-9} seconds, before the next photon comes along, 0.3 seconds later. Furthermore, in the third experiment the two detectors never clicked at the same time, a result ruling out the possibility that the incoming photons could have divided in two and passed through both slits at the same time. In the third experiment, the time between clicks was about 0.3 seconds.

So we have a contradiction! The fourth experiment requires that two waves of light pass almost simultaneously through the two slits, so that they can overlap on their way to the screen and produce the observed interference pattern. The third experiment shows that two photons never pass through the two slits at the same time; each photon passes through one slit at a time, and the resulting pattern of light is in accordance with such a picture: the sum of two single-slit patterns.

The contradiction between the third and fourth experiments is the enigma of the wave-particle duality of light and, as we will show, the duality of all nature. When we don't check to see which slit each photon goes through, each photon behaves as if it went through both slits at the same time, as a spread out wave would do. When we do check, each photon goes through *either* one slit *or* the other and behaves as a particle. Light behaves sometimes as a wave and sometimes as a particle. Astoundingly, and against all common sense, the behavior that occurs in a given experiment depends on what the experimenter chooses to measure. Evidently, the observer, and the knowledge sought by the observer, play some kind of fundamental role in the properties of the thing observed. The observer is somehow part of the system. These results call into question the long-held notion of an external reality, outside and independent of the observer. There is nothing more profound and disturbing in all of physics.

These, then, are the two enigmas of the quantum world: the wave-particle duality of nature and the strange role played by the observer. If you find these results impossible to fathom, you are in excellent company. Quantum effects continue to baffle the best physicists in the world.

2. Experimental Confirmation of the Double-Slit Experiment for Electrons

As we will discuss in section F–5, matter as well as light behaves both like a particle and like a wave. In particular, electrons sometimes behave like waves. In 1954, the German physicists G. Möllenstedt and H. Düker reported the first double-slit experiment to use electrons. [Their article appeared in *Naturwissenschaften*, vol. 42, p. 41 (1954); a similar experiment, using multiple slits, was reported by Claus Jönsson in *Zeitschrift für Physik*, vol. 161, p. 454 (1961); the latter article is partly translated in the *American Journal of Physics*, vol. 42, p. 4 (1974).] The experimental setup was just as shown in Fig. IV–6d, except for an incoming beam of electrons instead of an incoming beam of light. The electrons passed through slits and were detected at a screen. Möllenstedt and Düker found that when both slits were open but with no intermediate detectors, as in Fig. IV–6d, the electrons produced an interference pattern, like that shown in the right side of Fig. IV–3b. The electrons did not produce a pattern that was the sum of the patterns for each slit by itself. The electrons behaved as if they were overlapping waves rather than discrete particles.

D. THE ROLE OF THE OBSERVER AND THE NATURE OF REALITY

1. The Split Between Observer and Observed in Prequantum Physics

The idea of a clear distinction between the observer and the observed was embedded in prequantum science. Since the time of Galileo, in the seventeenth century, scientists had proceeded on the assumption that they could be passive observers of nature, without disturbing what they were looking at. Physicists believed that they could time the period of a swinging pendulum without changing its motion. Chemists believed they could measure the rate at which coal burned in air without altering that rate. Naturalists believed they could quietly listen to a sparrow without dictating its song. Without question, all scientists assumed that they could put a box around their subject and peer into that box. The scientist was on the outside, and the pendulum or coal or sparrow was on the inside. "The belief in an external world independent of the perceiving subject [observer] is the basis of all natural science," said Albert Einstein [Einstein in *James Clerk Maxwell: A Commemoration Volume* (Cambridge University Press: Cambridge, England, 1931)].

2. Interpretations of the Wave-Particle Duality of Nature

Besides the double-slit experiment, many other twentieth-century experiments have shown that matter and energy behave sometimes as waves and sometimes as particles, depending on what the experimenter chooses to measure and how she sets up the experiment. Such results naturally raise the questions: What is the nature of the matter or energy *before* an observation is made? In what form does a photon or electron exist *between* its emission and its detection? Is it half wave and half particle? If so, what does that mean? What is the role of the observer? Can the observer be separated from the observed? These are largely philosophical, not scientific questions. Science can deal only with measurements. Scientists can record whether a photon passes through one hole or another, and scientists can record where photons land on a photographic plate. Scientists can make theories to predict the result of a measurement. But the question of where and in what form a photon—or any other type of matter or energy—exists *between measurements of it* lies outside science.

Nevertheless, physicists, and indeed great physicists, have not refrained from speculation. There are two major schools of speculation. The first is called the "Copenhagen interpretation of quantum physics," named after the city in Denmark where the atomic physicist Niels Bohr (1885–1962) worked and attracted other brilliant physicists of the day. The Copenhagen interpretation of quantum physics, developed by Bohr and others in the 1920s, holds that prior to the measurement of an object, the object has no definite physical existence. Prior to a measurement, an object's existence and properties can be described only by various probabilities. A photon traveling toward the two slits in a screen, for example, would have some probability of being in one place and a different probability of being in another place. In the absence of a measurement, we cannot say exactly where the photon is; all we can say is that there is a 10% probability of its being in one place, a 25% percent probability of its being in another, and so on. At the moment the photon is measured—by a photographic plate, for example—it suddenly comes into existence, at a particular location, and the probability of its being in that location jumps to 100%.

The other major school of thought is called the "many-worlds interpretation of quantum physics" and was developed by the American physicists John Wheeler and H. Everett in the 1950s. In this interpretation, an object *does* have a physical existence prior to being measured. In fact, the object exists in all of its possible conditions and locations. *Each of these different existences occurs in a separate world.* Every time an observer makes a measurement of a photon and finds it to be at a particular place, the reality, or world, of that observer branches off from the other worlds and follows a track in which that particular photon and that particular observer have specific locations and properties. The situation is shown in a highly schematic and simplified form in Fig. IV–7. Since there are a fantastically large number of possible conditions for each particle, the many-worlds interpretation of quantum physics invokes a countless number of parallel worlds, all with real physical existence.

Both the Copenhagen interpretation and the many-worlds interpretation of quantum physics may seem bizarre, but so are the observed phenomena that they attempt to explain. In both interpretations, the observer plays a strong role in helping to shape reality: in the first, the observer creates reality; in the second, the observer causes many co-existing realities to split off from each other.

3. Berkeley's Views on External Reality

Questions of the nature of external reality and of the interplay between the observer and the observed have long been a subject of philosophical debate. In the twentieth century, for example, the group of modern philosophers known as existentialists have focussed on what determines human existence. The existentialists—who include the German philosophers Martin Heidegger and

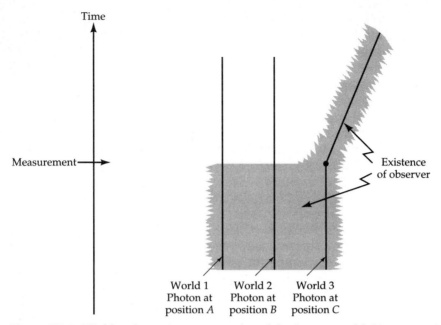

Figure IV–7: Highly schematic representation of the "many-worlds" interpretation of quantum physics. The observer exists in the shaded regions.

Karl Jaspers and the French philosophers and writers Jean-Paul Sartre and Albert Camus—hold that human beings are not detached observers of the world but exist "in the world." Human beings, the observers, are defined by their interaction with the world and have no existence independent of this interaction. This belief is the flip side of the statement that the properties of electrons and photons—the objects of the world—are determined by the act of their observation, that is, their interaction with an observer.

In the past, philosophers have often asked whether the observed objects of the world have an existence independent of our observation of them. Here, we will excerpt from the writings of the Irish philosopher and bishop George Berkeley (1685–1753). The passage below comes from Berkeley's *Three Dialogues*, which consists of a fictional debate between two men: Hylas and Philonous. Hylas believes that the objects we observe have a real existence outside of the human mind, while Philonous holds that only the observations themselves are real. Thus, according to Philonous, there is no reality external to the observer. Philonous repeatedly refers to "sensible" things, by which he means things that can be seen, heard, smelled, tasted, and touched—in other words, things that can affect the sense organs. Here is a representative passage:

PHIL: This point then is agreed between us—that sensible things are those only which are immediately perceived by sense. You will further inform me whether we immediately perceive by sight anything besides light and colors and figures; or by hearing, anything but sound; by the palate, anything besides tastes; by the smell, besides odors; or by the touch, more than tangible qualities.

HYL: We do not.

PHIL: It seems, therefore, that if you take away all sensible qualitites, there remains nothing sensible?

HYL: I grant it.

PHIL: Sensible things therefore are nothing else but so many sensible qualities or combinations of sensible qualities?

HYL: Nothing else.

PHIL: Heat is then a sensible thing?

HYL: Certainly.

PHIL: Does the reality of sensible things consist in being perceived, or is it something distinct from their being perceived, and that bears no relation to the mind?

HYL: To exist is one thing, and to be perceived is another.

PHIL: I speak with regard to sensible things only; and of these I ask, whether by their real existence you mean a subsistence exterior to the mind and distinct from their being perceived?

HYL: I mean a real absolute being, distinct from and without any relation to their being perceived.

PHIL: Heat therefore, if it be allowed a real being, must exist without the mind?

HYL: It must.

PHIL: Tell me, Hylas, is this real existence equally compatible to all degrees of heat, which we perceive, or is there any reason why we should attribute it to some and deny it to others? And if there be, pray let me know that reason.

HYL: Whatever degree of heat we perceive by sense, we may be sure that the same exists in the object that occasions [produces] it.

PHIL: What! The greatest as well as the least?

HYL: I tell you, the reason is plainly the same in respect of both: they are both perceived by sense; nay, the greater degree of heat is more sensibly perceived; and consequently if there is any difference, we are more certain of its real existence than we can be of the reality of a lesser degree.

PHIL: But is not the most vehement and intense degree of heat a very great pain?

HYL: No one can deny it.

PHIL: And is any unperceiving thing capable of pain or pleasure?

HYL: No, certainly.

PHIL: Is your material substance a senseless being [a being without senses] or a being endowed with sense and perception?

HYL: It is senseless, without doubt.

PHIL: It cannot, therefore, be the subject of pain?

HYL: By no means.

PHIL: Nor, consequently, of the greatest heat perceived by sense, since you acknowledge this to be no small pain?

HYL: I grant it.

PHIL: What shall we say then of your external object: is it a material substance, or no?

HYL: It is a material substance with the sensible qualities inhering in it [existing as part of it].

PHIL: How then can a great heat exist in it, since you own it cannot in a material substance? I desire you would clear this point.

HYL: Hold, Philonous, I fear I was out [mistaken] in yielding [agreeing] intense heat to be a pain. It should seem rather that pain is something distinct from heat, and the consequence or effect of it.

PHIL: Upon putting your hand near the fire, do you perceive one simple uniform sensation or two distinct sensations?

HYL: But one simple sensation.

PHIL: Is not the heat immediately perceived?

HYL: It is.

PHIL: And the pain?

HYL: True.

[George Berkeley, *Three Dialogues Between Hylas and Philonous* (1713) (Bobbs-Merrill: New York, 1954), pp. 13–15]

In the last few exchanges, Philonous is saying that it is hard to distinguish between a sensory sensation and the cause of that sensation. The quality and nature of heat cannot be untangled from the ability to make something hot, and so on. Notice that Philonous has cleverly limited his discussion to objects that are perceived and sensed by human beings. As such, his arguments are hard to refute, and he makes a fool of Hylas. However, we could legitimately ask about the reality of things that are not perceived by human beings. Of course, if we have no interaction with an object whatsoever, we cannot *prove* the object exists. Yet, it does seem difficult to imagine that no objects exist except those that we have observed, no galaxies exist except those we have sighted through our telescopes, no flowers exist except those we have personally smelled and enjoyed. Philonous's idea is poetic but hard to swallow. On the other hand, the debate shows Berkeley's strong interest in the fundamental relationship between objects and their observers.

Discussion Questions IV–1

Discuss the logic of Philonous's argument that objects do not have an existence independent of their perception. Do you see any flaws in his logic? Is Philonous justified in disregarding all things that are not "sensible" things?

In Philonous's view, does the observer interact with the observed, or simply observe it? Is Philonous saying that there is no way we can *prove* the existence of an object outside our sense perception, or is he saying that such an unperceived object doesn't exist in the first place? Do you think Berkeley is expressing his own view here and, if so, which character does he identify with?

Discussion Activity IV–1

Divide the class into two groups, one representing Hylas and one Philonous, and continue the dialogue.

Discussion Questions IV–2

If the observer is part of the system being observed, then how can science claim an objective view of the world? Do scientists always find what they are looking for? Think of examples from earlier chapters and discuss.

E. QUANTUM PHYSICS AND LANGUAGE

The wave-particle duality of nature raises questions about our perceptions of the world, and it also raises questions about our attempt to describe the world, that is, our language. The quotation from Niels Bohr at the beginning of this chapter shows the struggle to come to terms with the meaning of quantum physics. The following quotation from Werner Heisenberg (1901–1976), another giant in the development of quantum physics, bears upon that struggle.

> Any experiment in physics, whether it refers to phenomena of daily life or to atomic events, is to be described in the terms of classical [nineteenth-century] physics. The concepts of classical physics form the language by which we describe the arrangement of our experiments and state the results. We cannot and should not replace these concepts by any others. [Werner Heisenberg, *Physics and Philosophy* (Harper: New York, 1958), p. 44]

The language of ordinary perceptions and classical physics that Bohr and Heisenberg refer to includes such words as "particle," "wave," "object," "position," "motion," "observer," "measurement," and "causality." These words have been used for centuries to indicate our understanding of the nature of things and our relationship to those things. In part, language is an attempt to convey the visual pictures we carry around in our heads. Yet, quantum physics has shown us that these visual pictures can no longer account for the phenomena of nature as recorded by the instruments of modern science. We have no intuition for the behavior of photons in the double-slit experiment; we have no visual picture that can account for how a photon or an electron seems to be in two places at once.

Discussion Questions IV–3

Do Bohr and Heisenberg suggest that our current language is inadequate to describe the real world? Why do you think Heisenberg says we should continue to speak in terms of classical concepts?

Discussion Questions IV–4

How is a language born and what is the role of experience in a language? Is there a way to enlarge our language and concepts beyond what we experience with our bodies? Is so, would this enlarged language be useful?

F. THE HEISENBERG UNCERTAINTY PRINCIPLE AND THE DEMISE OF DETERMINISM IN SCIENCE

We will see in this section how the wave-particle duality leads to a fundamental change in our ability to make predictions about the world. In prequantum physics, we could, in principle, predict the trajectory of an individual electron with complete certainty. Quantum phenomena cause a basic uncertainty in the path of a particle. We can make predictions only about how a large number of electrons or photons or atoms will behave. If there are 10^{30} electrons in motion in a region of space, we can predict with good accuracy the average trajectory, but we cannot predict the motion of an individual electron. Predictions of nature thus become probabilistic, or statistical. No longer do the individual pieces of nature obey fully deterministic laws. This uncertainty and indeterminancy—which seem to stem from an intrinsic property of nature rather than from inadequate knowledge—disturbed Einstein so much that he could never fully accept quantum physics.

We will begin with a discussion of the determinism in prequantum physics, then state the basic idea of the Heisenberg uncertainty principle, and end with a detailed derivation of the quantitative uncertainty relations, Eqs. (IV–28).

1. Determinism in Prequantum Physics

Beginning with the work of Issac Newton and reaching a grand climax with the formulations of the French physicist and mathematician Pierre Simon Laplace (1749–1827), scientists arrived at a completely deterministic view of the world. According to this view, the future position of a planet or a speck of dust could be completely predicted once its present conditions were known. The Laplacian world, as it is sometimes called, was nothing but a machine. A snapshot of the machine at one instant, plus a knowledge of its workings, would determine the future for all time.

To illustrate the deterministic nature of prequantum physics, let us consider the motion of a ball thrown upward. The ball will have some initial height y_i, and some initial upward speed, v_i. (As mentioned earlier, we will denote all vertical heights by y in this chapter, to avoid confusion with the traditional symbol for Planck's constant, h.)

As the ball rises upward, gravity will tug downward on it with a gravitational acceleration of $g = 9.8$ meters per second per second, as we discussed in Chapter I. We want to predict the height of the ball after an elapsed time interval Δt. Let y_f and v_f be the final height and speed of the ball, after a time interval of Δt. As we learned in Eq. (I–5b), the change in height of the ball after a time Δt is

$$\Delta y = y_f - y_i = \left(\frac{v_i + v_f}{2} \right) \Delta t, \qquad \text{(IV–7a)}$$

and the change in speed from the gravitational acceleration is, from Eq. (I–5a),

$$\Delta v = v_f - v_i = -g\, \Delta t. \qquad \text{(IV–7b)}$$

Here, there is a minus sign in front of the g in Eq. (IV–7b) because the gravitational acceleration is downward and thus decreases the upward speed v, so that v_f is less than v_i. Also, the final height y_f is greater than the initial height y_i in Eq. (IV–7a) as long as the average upward speed $(v_i + v_f)/2$ is positive. From Eq. (IV–7b), we conclude that

$$v_f = v_i - g\, \Delta t. \qquad \text{(IV–7c)}$$

Substituting this value of v_f into Eq. (IV–7a) and solving for y_f, we obtain

$$y_f = y_i + \left(\frac{v_i + v_i - g\,\Delta t}{2} \right) \Delta t = y_i + v_i\, \Delta t - \frac{1}{2} g(\Delta t)^2. \qquad \text{(IV–8)}$$

Equation (IV–8) gives the height of the ball after an elapsed time of Δt, given the ball's initial height y_i and initial speed v_i. It is a predictive equation in that it tells us where the ball will be at any later time, given its position and speed at the current time. It is deterministic.

Most equations of physics are similar to Eq. (IV–8). The future position of a particle is completely determined if three things are known: the initial position of the particle [y_i in Eq. (IV–8)], the initial speed of the particle [v_i in Eq.

(IV–8)], and the acceleration acting on the particle [g, resulting from gravity, in Eq. (IV–8)]. Even if you have trouble deriving Eq. (IV–8), you can take it as a model for a deterministic equation in physics.

Now, according to prequantum physics, the intitial conditions of a particle—that is, a particle's initial position and initial speed—could in principle be determined with infinite accuracy. We say "in principle" because any *actual* measuring device has certain limitations and cannot measure anything with 100% accuracy. But, prior to the understanding of quantum physics, scientists believed that there was no limit on how accurately the initial conditions of a particle might be measured. Measuring devices could be continually improved and the initial conditions of a particle determined more and more accurately. Accordingly, the future trajectory of a particle could be predicted more and more accurately. This advance in accuracy could continue indefinitely, until the future trajectory of a particle could be predicted with as much accuracy and certainty as anyone wanted.

In 1927, the German physicist Werner Heisenberg, a theorist, showed that quantum physics and the wave-particle duality of nature forbid the precise measurement of the initial conditions of a particle, or anything else. If the initial conditions cannot be measured precisely, then the future conditions cannot be predicted precisely. Heisenberg's result is not a statement about our inability to construct good measuring devices. It is a statement about an intrinsic property of nature. No matter how good our measuring devices are, the behavior of nature is such that we cannot measure things to better than a certain accuracy. Nature has an essential indeterminacy. Nature can be pinned down only so far and no farther. Heisenberg's result, called the Heisenberg uncertainty principle, is one of the most important and famous discoveries of all science.

2. The Basic Idea of the Principle

The basic idea behind the Heisenberg uncertainty principle is that all matter and energy behave partly like a wave and partly like a particle. For the moment, we will use the word "particle" to describe a bit of matter, such as an electron or a photon. Let's measure the position and motion of a particle at the moment it passes through a small hole, such as a pinhole in a window shade. It is a fact of nature that the particle acts partly like a wave. When a wave passes through a narrow hole, it spreads out in a series of outgoing ripples, as we saw in Fig. IV–3a. Analogously, a water wave moving past a branch on a pond breaks up into many ripples and goes outward in all directions.

As a result of such spreading out "downstream" of the hole, we cannot tell exactly the direction of any particular particle when it emerges from the hole. Thus, we cannot accurately specify the initial conditions of the particle at the hole. Consequently, we cannot accurately determine the particle's future trajectory. The amount of spreading out gets bigger the smaller the hole. If we make the hole very small, so that we know very accurately the initial *position* of the particle, the uncertainty in the particle's initial *direction* becomes very large. If we make the hole very large, in order to greatly reduce the spreading out effect, then the uncertainty in initial position becomes very large, because the particle could begin its trajectory anywhere inside the area enclosed by the hole. So, the wave-particle duality leads to an unavoidable tradeoff: we can accurately measure the initial position *or* the initial motion of a particle, but not both. Unfortunately, an accurate determination of the future trajectory requires that we accurately specify both. We are stuck with an unavoidable uncertainty.

We are ready to make all this quantitative. The derivation will require some intermediate steps in order to sharpen our understanding of waves and particles, and the math will require some patience. We will follow this plan: (1) First, we will consider how waves interfere and spread out. This section will be the hardest part mathematically. We will conclude this section with a quantitative formula for the angle of spreading of light through a hole in terms of the wavelength of light and the size of the hole, Eq. (IV–17). (2) Second, we will reinterpret our results of the spreading of light in terms of a particle-like picture of light. In this picture, photons that initially have speeds only in the horizontal direction gain an additional speed in the vertical and sideways directions after passing through a hole; it is these added components of speed that constitute the "spreading" of a group of photons. We will intro- duce a new concept, momentum, which is closely related to speed, and describe the photon spreading in terms of the photon momenta in the vertical and horizontal directions. (3) Next, we will derive a relation between momen- tum, wavelength, and Planck's constant, Eq. (IV–24). (4) Finally, by combining this relation with our previous results, we will obtain the Heisenberg uncer- tainty principle, Eq. (IV–28). This last equation quantitatively expresses the relation between the uncertainty in position and the uncertainty in momen- tum of a photon or any other particle.

If you cannot find the stamina to tackle the whole calculation, you can read the discussion of momentum surrounding Eqs. (IV–19), (IV–21) and (IV–22), and then leap to the final result, Eq. (IV–28a).

3. Spreading Out of Light as a Wave

We will begin with light. When light passes through a small hole, it spreads out. It is difficult to observe such spreading of light with the naked eye, because the wavelength of visible light is so small, but a similar effect can be seen with water waves, where the wavelengths are usually much larger. If water in a pond flows past an upright stick, you can see ripples spreading out downstream of the stick.

In this section, we will calculate the angle of spreading for light passing through a small hole. That angle will depend on the size of the hole and the wavelength of the light.

The spreading effect is illustrated in Fig. IV–8a. Incoming light waves, all moving parallel to each other, pass through a hole in a shade. From each point in the hole, outgoing waves of light emerge in all directions. Figure IV–8a is a snapshot; it shows the pattern of outgoing waves at a moment of time. At a later instant of time the pattern will have moved to the right, since the waves are moving through space to the right.

If a screen is placed in front of the outgoing light, a pattern of light and dark will appear on the screen. This pattern is produced by the interference of

Figure IV–8: (*a*) Spreading out of light waves after emerging through a hole in a shade. (*b*) Analysis of the interference of three waves, emerging from the top, middle, and bottom of the hole. All three waves converge at point *A* on the screen. The point *M* on the screen, the midway point, is directly opposite the middle of the hole.

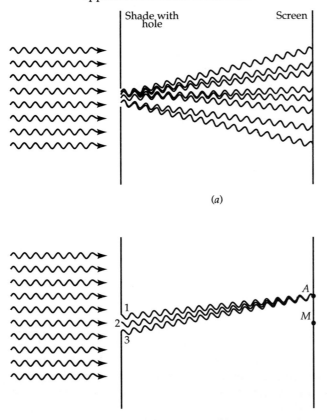

(*a*)

(*b*)

different waves striking the screen. Just as we saw in section A–5 and Fig. IV–3 how light waves emerging from two slits could overlap and interfere with each other, so light waves emerging from a single slit or hole can overlap and interfere with each other.

Each point of the screen receives many different light waves. Points on the screen at which the crests of the arriving light waves line up, so that they constructively interfere, are bright. Points on the screen at which the arriving light waves destructively interfere, canceling each other out, are dark.

Figure IV–8b, for example, illustrates the situation for point A on the screen. Many different light waves from the hole in the shade arrive at point A. For simplicity, we have drawn only three of those waves: the wave coming from the top of the hole, labeled 1, the wave from the middle of the hole, labeled 2, and the wave from the bottom of the hole, labeled 3. The intensity of light at point A depends on how the crests and troughs of waves 1, 2, and 3 line up with each other at that point.

Figure IV–8b is another snaphot; it shows the situation at a single moment of time. At a later moment, the patterns of waves 1, 2, and 3 will shift to the right, since those waves are in motion. However, if all the waves have the same frequency, then the way in which they overlap at point A will not change in time because waves of the same frequency shift their patterns through space at precisely the same rate. Thus, if at one moment of time two waves reach point A when both are at their low points (their troughs), then at a later moment of time, when one wave is at its high point, the other will be as well. If two waves enhance each other at point A at one moment of time, they will enhance each other at any later moment; if two waves cancel each other at point A at one moment, they will always cancel each other there. For these reasons, the interference of waves of a single frequency (or wavelength) can be analyzed with snapshots.

There will be a bright spot on the screen directly in front of the middle of the hole. Our ultimate question is: how big is that bright spot? The size of the bright spot is directly related to how much the incoming light rays have spread out in emerging from the hole. First we need to know the conditions for the waves to constructively interfere, causing a bright spot.

Consider the three waves arriving at point A on the screen in Fig. IV–8b, and focus on the two extreme waves: wave 1 from the top of the hole and wave 3 from the bottom. *These waves travel different distances to get to point A.* Wave 3 must travel farther. Only at point M, lying directly opposite the middle of the hole, will waves 1 and 3 travel the same distance to get to the screen. For any point A above M, the number of crests of wave 1 between the hole and point A will be fewer than the number of crests of wave 3.

If wave 3 travels a whole number of wavelengths more than wave 1 to get to point A—that is, if a snapshot shows that wave 3 has one *complete* crest and trough more than wave 1, or two *complete* crests and troughs, or more—

then the two waves will be exactly lined up at point A and will completely reinforce each other there. Figure IV–8c illustrates such a situation where the two waves constructively interfere at point A. The distance wave 1 travels to point A is 12 wavelengths; the distance wave 3 travels is 13 wavelengths. Both waves are at their troughs at point A.

If, on the other hand, wave 3 travels a fractional number of wavelengths more than wave 1 to get to point A—say, 1.3 wavelengths more or 2.5 wave-

Figure IV–8: (*c*) Analysis of the two extreme waves, when their difference in travel distance from the hole to point A is exactly 1 wavelength. (*d*) Analysis of the two extreme waves, when their difference in travel distance from the hole to point A is exactly half a wavelength.

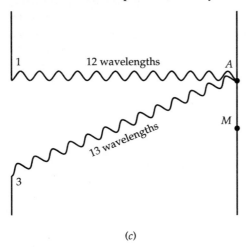

(*c*)

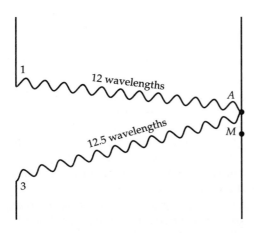

(*d*)

lengths more—then the two waves will not completely reinforce each other. In Fig. IV–8d, for example, wave 1 travels 12 wavelenths to get to point A and wave 3 travels 12.5 wavelengths. At point A, wave 1 is at its trough and wave 3 is at its crest. Therefore, the two waves cancel each other out completely at point A.

If point A is sufficiently close to the midpoint of the screen, point M, then the difference in length traveled by waves 1 and 3 in reaching point A will be smaller than half a wavelength. (Remember, at the midpoint M the distance traveled by waves 1 and 3 is exactly the same.) In such a case, the crests of waves 1 and 3 will be almost lined up with each other at point A. Since waves 1 and 3 represent the extremes in distance traveled, all intermediate waves, such as wave 2 in Fig. IV–8b, will be even more closely lined up with each other than waves 1 and 3 when they strike the screen. Thus, point A will receive a lot of light and it will be bright.

As point A moves farther and farther from the midpoint M, the difference in distance traveled by waves 1 and 3 increases. *At some critical point, this difference is half a wavelength.* (Point A is located at the critical point in Fig. IV–8d.) At this critical point, waves 1 and 3 cancel each other out. The light is still bright out to the critical point because only the two extreme light rays, out of an enormous number, have canceled each other out. However, when point A is farther from point M than this critical point, intermediate waves (originating from between the top and bottom of the hole) also begin canceling each other out. Inside the critical point, such intermediate waves were almost lined up with each other when they reached the screen and thus produced a lot of light on the screen. But outside the critical point, the intermediate waves begin canceling each other out. The resulting light at the screen outside the critical point begins to dim, because fewer uncanceled rays are reaching such regions. To summarize, the screen will be bright in a circle around point M, directly in front of the middle of the hole, and dim beyond that circle. *The edge of the circle of light is the "critical distance" from M, such that waves 1 and 3 travel distances differing by half a wavelength to get to this edge.*

The situation is illustrated in Fig. IV–9. Figure IV–9a shows the different waves arriving at the screen. Figure IV–9b shows a front view of the screen. Point M, directly in front of the middle of the hole in the shade, lies at the center of the circle of light on the screen. The radius of the circle of light is x. We want to determine x, which we can consider to be the distance over which the incoming light spreads out.

Now, we need to do some algebra to quantify what we have just said. *From the previous discussion, the condition that determines x is that the distances to the edge of the circle of light traveled by the two extreme waves, waves 1 and 3, must differ by half a wavelength.* The edge of the circle is a distance x away from M. Let λ stand for the wavelength of light, l_1 for the distance wave 1 travels from the shade to a point x away from M, and l_3 for the distance wave 3 travels to the same point. Then the required condition that determines x is

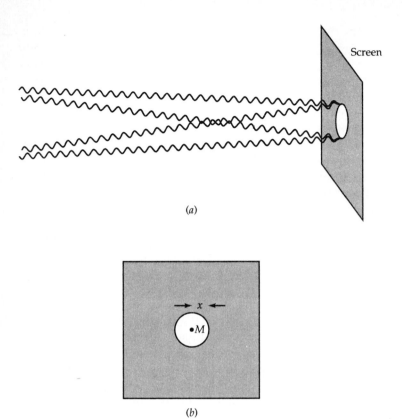

(a)

(b)

Figure IV–9: (a) Illumination of a screen by light passing through a hole in a shade. (b) A close-up, front view of the illuminated screen. Point M is the same as in Fig. IV–8. The radius of the circle of light is x.

$$l_3 - l_1 = \tfrac{1}{2}\lambda. \tag{IV–9}$$

Figure IV–10a shows the geometry. Here the width of the hole in the shade is denoted by a and the distance between the shade and the screen is d. The dashed line extends from the middle of the hole to the midpoint of the screen M, and the circle of light will be centered on M. To determine l_1 and l_3 in terms of x, a, and d, note that two right triangles may be constructed, one having l_1 as its hypothenuse and one having l_3 as its hypothenuse. (See Appendix A for a review of the relevant geometry.)

The geometry of these two right triangles is analyzed in Figs. IV–10b and c. The triangle involving l_1, shown in Fig. IV–10b, has legs d and $x - a/2$. Thus, by the Pythagorean theorem,

$$l_3^2 = d^2 + \left(x + \frac{a}{2}\right)^2. \tag{IV–10a}$$

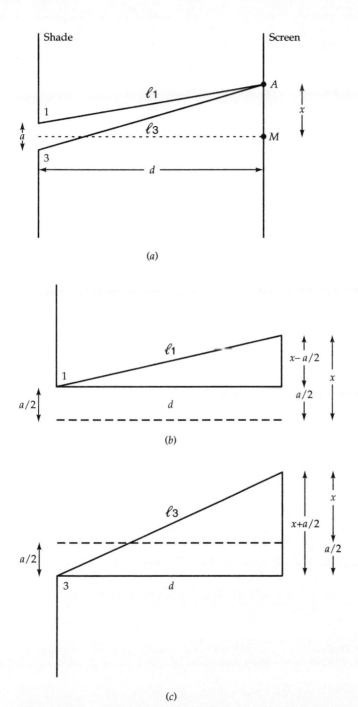

Figure IV–10: Geometrical constructions used in analyzing the distances traveled by waves 1 and 3 from the hole to the screen.

The triangle involving l_3, shown in Fig. IV–10c, has legs d and $x + a/2$. By the Pythagorean theorem,

$$l_3^2 = d^2 + \left(x + \frac{a}{2} \right)^2. \tag{IV–10b}$$

What we want to calculate, in order to substitute into Eq. (IV–9), is $l_3 - l_1$. From simple algebra, we know that $(l_3 - l_1)(l_3 + l_1) = l_3^2 - l_1^2$. Solving this equation for $l_3 - l_1$, we get

$$l_3 - l_1 = \frac{l_3^2 - l_1^2}{l_3 + l_1}. \tag{IV–11}$$

Now we're going to make an approximation. Let us assume that d is much larger than either a or x. Then, Eqs. (IV–10) show that both l_3 and l_1 are almost equal to d. So, in the denominator of Eq. (IV–11), we can substitute the approximation

$$l_1 + l_3 \approx 2d. \tag{IV–12}$$

Here, as before, we use the symbol $\approx$ to indicate an approximate equality. Substituting Eq. (IV–12) into Eq. (IV–11), we get

$$l_3 - l_1 \approx \frac{l_3^2 - l_1^2}{2d}. \tag{IV–13}$$

Note that even though l_3 and l_1 are almost equal in our approximation, it is important to keep track of their small difference in the numerator of Eq. (IV–13). Otherwise, we would get zero and the entire effect of interference would disappear. But in *adding* l_1 and l_3, as in the denominator of Eq. (IV–11), their small difference can be neglected.

Now, substituting Eqs. (IV–10) into Eq. (IV–13), we obtain

$$l_3 - l_1 \approx \frac{(d^2 + x^2 + xa + a^2/4) - (d^2 + x^2 - xa + a^2/4)}{2d},$$

or, making the subtractions,

$$l_3 - l_1 \approx \frac{xa}{d}. \tag{IV–14}$$

Finally, combining Eqs. (IV–14) and (IV–9), we get $xa/d \approx \lambda/2$, or

$$x \approx \frac{\lambda d}{2a}. \tag{IV–15}$$

We have kept the $\approx$ sign rather than the $=$ sign to remind ourselves that Eq. (IV–15) is an approximation, valid only when d is much larger than either a or x. From Eq. (IV–15), we can see that this restriction is equivalent to the conditions that both a and x are much larger than λ.

The distance x, which marks the radius of the circle of light on the screen, measures how much the light has spread out in traveling from the opening in the shade to the screen, a distance d away. It is more convenient to measure the spreading by the *angle* at which the light rays fan out from the opening, as illustrated in Fig. IV–11. Denote that angle by θ. If we draw a line from the middle of the opening in the shade to the edge of the circle of light on the screen, a distance x above M, then θ is the angle between that diagonal line and the dashed line from the middle of the opening to point M. Since θ is one angle of a right triangle, two of whose legs are x and d, θ is completely determined by the ratio x/d. (See Appendix A for a brief review of the relevant geometry and trigonometry.) For example, if $x/d = 1$, $\theta = 45°$; if $x/d = 2$, $\theta = 63.4°$. A shorthand notation for the relation between θ and x/d is

$$\tan \theta = \frac{x}{d}. \tag{IV–16}$$

Using Eq. (IV–16) and an electronic calculator with the tan (tangent) function, you can easily compute the value of x/d corresponding to each value of θ. (And with the inverse tangent function, you can compute the value of θ corresponding to each value of x/d.)

Figure IV–11: Representation of the effective angle by which light spreads after emerging from the hole.

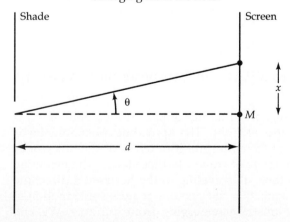

Substituting Eq. (IV–15) into Eq. (IV–16), we get

$$\tan \theta \approx \frac{\lambda}{2a}. \qquad \qquad \text{(IV–17)}$$

Equation (IV–17) is our final result. It gives the angle by which light spreads out after emerging from a hole. Notice that the angle depends only on the ratio of the wavelength of incoming light to twice the diameter of the hole, $\lambda/2a$. If we make the wavelength of light larger, the spreading angle θ becomes larger. If we make the hole larger, the spreading angle becomes smaller. These results make sense. The spreading out effect is caused by the wavelike nature of light. The smaller the wavelength of light relative to openings the light ray encounters, the less light looks like a wave.

Problem IV–4: Spreading Through a Hole

Suppose that a very straight beam of orange light passes through a hole of diameter 2×10^{-4} centimeters. What is the angular spread of the emerging beam of light?

Solution: From Problem IV–3, the wavelength of orange light is $\lambda = 6 \times 10^{-5}$ centimeters. We are given that the diameter of the hole is $a = 2 \times 10^{-4}$ centimeters. Substituting these values for λ and a into Eq. (IV–17), we get

$$\tan \theta \approx \frac{6 \times 10^{-5} \text{ cm}}{4 \times 10^{-4} \text{ cm}} = 0.15,$$

giving for θ

$$\theta = 8.5°.$$

---

4. Spreading Out of Light as a Particle: Components of Velocity and Momentum

In this section, we will discuss the spreading out of light from a hole in terms of the particle, or photon, picture of light. The spreading phenomenon is exactly the same. We simply want to reinterpret it in terms of the trajectories of photons rather than the interference of waves. In these terms, the incoming light consists of a stream of photons, all traveling in the horizontal direction. The spreading out of the light means that the photons acquire some motion in the vertical and left-right directions after emerging from the hole. We will arrive at the Heisenberg uncerainty principle by combining the particle-like picture of the photon from this section with the wavelike picture of the previous section.

The particle-like view is illustrated in Fig. IV–12. Here, each little circle represents a photon. Arrows have been attached to some of the photons to show their direction of motion. Some of the photons will continue traveling in a completely horizontal direction after passing through the hole; others will have varying degrees of motion in the vertical and left-right directions in addition to their horizontal motion. Figures IV–10, IV–11, and IV–12 show only the vertical and horizontal directions. To see the left-right direction as well, we need to twist the diagram a little, as is done in Fig. IV–13a. Note that both the vertical and the left-right direction are perpendicular to the horizontal direction. We will continue denoting distances in the vertical direction by y; we will denote distances in the left-right direction by z.

Suppose we look at the photons that change direction the most, that is, those photons that acquire the largest vertical or left-right motions after passing through the hole. Such extreme photons spread out the most and land at the edge of the circular bright spot on the screen. All other photons spread out less and land somewhere in the interior of the bright spot on the screen. The angle of motion of the extreme photons is the same as the angle θ shown in Fig. IV–11. It is the spreading angle.

We now want to relate the spreading angle to the photon speeds in the horizontal, vertical, and left-right directions. We will denote the speeds in these three directions by v_h, v_y, and v_z, respectively. (The net speed of the photon is c, as we have discussed many times, but the photon can have lesser speeds along particular directions, just as a car traveling northwest at a net

Figure IV–12: Spreading of light upon emerging from a hole in a shade, with light represented by photons. The arrows indicate the direction of motion of the photons.

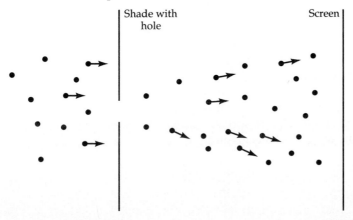

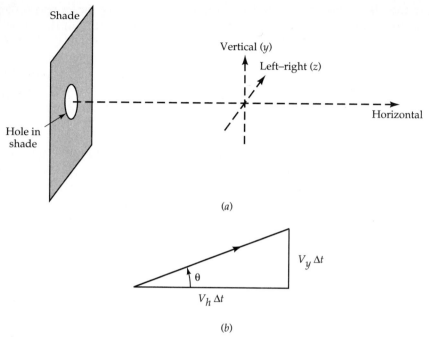

Figure IV–13: (*a*) Analysis of the spreading of light in terms of the speed of photons in different directions. Before passing through the hole, the photons travel only in the horizontal direction. After passing through the hole, they have speeds in the vertical and left-right directions as well. (*b*) Analysis of the spreading angle in the vertical direction. This analysis requires that we look at the photon speeds only in the vertical and horizontal directions.

speed of 100 kilometers is making good only 71 miles per hour in the northerly direction.) For simplicity, let's focus just on the motions in the horizontal and vertical directions. Since the hole in the shade is circular, whatever happens in the vertical direction (*y*) will happen exactly the same in the left-right direction (*z*). In a time interval Δt, a photon with vertical speed v_y goes a distance in the vertical direction $v_y \Delta t$. Likewise, it goes a horizontal distance $v_h \Delta t$ in the same time period. These two distances in the vertical and horizontal directions are shown in Fig. IV–13b. Thus, a right triangle is formed, with two legs equal to $v_y \Delta t$ and $v_h \Delta t$. The spreading angle θ lies opposite the leg of length $v_y \Delta t$, and so the tangent of θ is given by

$$\tan \theta = \frac{v_y \, \Delta t}{v_h \, \Delta t} = \frac{v_y}{v_h}. \tag{IV–18}$$

What we have done so far in this section is relate the spreading angle of the light to the vertical and horizontal speeds of the photons that are most deflected. We are almost finished. It will be useful to describe the motion of photons and other particles in terms of *momentum* instead of speed. Momentum is a quantity closely related to speed and figures into many of the laws of physics. Furthermore, the statement of our final result, the Heisenberg uncertainty principle, is much simpler in terms of momentum than speed. Let us denote momentum by p. The definition of the momentum p of a particle is

$$p = \frac{Ev}{c^2}, \tag{IV–19a}$$

where E is the energy of the particle, v is the speed of the particle, and c is the speed of light. Like speed, momentum can be broken up into vertical momentum, p_y, and horizontal momentum, p_h, by simply substituting the vertical and horizontal speeds into Eq. (IV–19a), that is,

$$p_y = \frac{Ev_y}{c^2}, \tag{IV–19b}$$

$$p_h = \frac{Ev_h}{c^2}. \tag{IV–19c}$$

Note that

$$\frac{p_y}{p_h} = \frac{Ev_y / c^2}{Ev_h / c^2} = \frac{v_y}{v_h}.$$

Thus, the ratio of vertical to horizontal speeds is the same as the ratio of vertical to horizontal momenta, and Eq. (IV–18) can be rewritten as

$$\tan \theta = \frac{p_y}{p_h}. \tag{IV–20}$$

We will soon discover that the Heisenberg uncertainty principle applies to matter and energy in all forms, not just to photons. Thus, it will be useful to apply our general definition of momentum, Eq. (IV–19a), to various kinds of particles, moving at both slow and fast speeds. For photons, the speed v is always c, so a photon's momentum is, using Eq. (IV–19a),

$$p = \frac{Ec}{c^2} = \frac{E}{c}. \tag{IV–21}$$

Equation (IV–21) actually gives the momentum of any particle traveling at the speed of light in terms of the particle's energy E. What about the momentum of a slowly moving particle? Recall from Chapter III, Eq. (III-13b), that for a particle moving slowly compared to the speed of light, the energy E is approximately

$$E \approx mc^2,$$

which is just the "rest energy." Substituting this value for E into Eq. (IV–19a), the momentum of a particle moving slowly compared to the speed of light is

$$p \approx mv. \tag{IV–22}$$

Thus, given the general definition for momentum, Eq. (IV–19a), we have two extreme cases: particles traveling at the speed of light, with momentum given by Eq. (IV–21), and particles traveling slowly compared to the speed of light, with momentum given by Eq. (IV–22). The Heisenberg uncertainty principle will apply to both extremes.

5. The de Broglie Relation

Before arriving at the Heisenberg uncertainty principle, we need one final and crucial result, discovered by the French physicist Louis de Broglie (1875–1960). If we combine Eq. (IV–21), expressing the relation between momentum and energy for a photon, and Eq. (IV–6), expressing the relation between a photon's energy and frequency, $E = h\nu$, we obtain

$$p = \frac{h\nu}{c}. \tag{IV–23}$$

Now, for light, we also have the result of Eq. (IV–2), relating wavelength λ to frequency ν, $\lambda = c/\nu$. Substituting $\nu = c/\lambda$ into Eq. (IV–23), we get

$$p = \frac{h}{\lambda}.$$

Finally, multiplying both sides by λ and dividing by p, we get

$$\lambda = \frac{h}{p}. \tag{IV–24}$$

Equation (IV–24) says that the wavelength of any photon equals Planck's constant, $h = 6.6 \times 10^{-34}$ joules $\times$ seconds, divided by the momentum of the photon. Note that Eq. (IV–24) is a hybrid. It combines the wavelike nature of light, embodied in the wavelength λ, with the particle-like nature of light, embodied in the photon momentum p.

In 1925, de Broglie proposed that Eq. (IV–24) should apply to *all* matter, not just to photons. Since matter appears to have both wavelike and particle-like properties, objects normally thought of as particles, like electrons, should have a wavelength corresponding to the wavelike half of their personality. De Broglie argued that any object with a momentum p, that is, any moving object, behaves partly like a wave and has a wavelength given by Eq. (IV–24). Thus, a beam of electrons or golf balls should exhibit interference phenomena just like photons. Such phenomena have indeed been observed, as mentioned in section C–2.

De Broglie, whose full name was Louis-Cesar-Victor-Maurice de Broglie, was a member of an old and venerable French family that had produced a long line of soldiers and politicians. De Broglie challenged his grandfather's attempts to make him a military man by building his own private laboratory in one of the rooms of the family mansion. There he carried out his research in physics. De Broglie received the Nobel prize in physics in 1929 for discovering the wavelike character of electrons. In his Nobel speech, delivered in Stockholm, Sweden, in 1929, he said,

> Some thirty years ago, Physics was divided into two camps. On the one
> hand there was the Physics of Matter, based on the concepts of corpus-
> cles [particles] and atoms which were assumed to obey the classical laws
> of Newtonian Mechanics; on the other hand there was the Physics of
> radiation, based on the idea of wave propagation in a hypothetical con-
> tinuous medium: the ether of Light and of electromagnetism. But these
> two systems of Physics could not remain alien to each other: an amalga-
> mation had to be effected . . . [and] by a brilliant piece of intuition
> Planck succeeded in doing so. Instead of assuming, as did the classical
> Wave Theory, that a light-source emits its radiation continuously,
> he assumed that it emits it in equal and finite quantities—in quanta.
> (p. 1049)

> I obtained the following general idea, in accordance with which I pur-
> sued my investigations—that it is necessary in the case of Matter, as well
> as of radiation generally and of Light in particular, to introduce the idea
> of the corpuscle [particle] and of the wave simultaneously. (p. 1051)

> We can no longer imagine the electron as being just a minute corpuscle
> of electricity: we must associate a wave with it. And this wave is not just
> a fiction: its length can be measured and its interferences calculated in
> advance. In fact, a whole group of phenomena was in this way predicted
> before being actually discovered. It is, therefore, on this idea of the dual-
> ism in Nature between waves and corpuscles, expressed in a more or less

abstract form, that the entire recent development of theoretical Physics has been built up, and that its immediate future development appears likely to be erected. (p. 1059) [Louis de Broglie, "The Undulatory Aspects of the Electron," Nobel Prize Address, in *The World of the Atom*, ed. H. A. Boorse and L. Motz, (Basic Books: New York, 1966), vol. II. pp. 1049–1059]

Problem IV–5: Electron Waves

An electron has a mass of about 10^{-30} kilograms. Consider an electron traveling at a speed of 2×10^6 meters per second, the typical speed of an electron in an atom. What is the wavelength of this electron?

Solution: We need only calculate the momentum p of the electron and then substitute into Eq. (IV–24). Since the speed of light is $c = 3 \times 10^8$ meters per second, the electron is traveling slowly compared with the speed of light. We can therefore use the approximation for its momentum given by Eq. (IV–22). Substituting in the numbers, we get

$$p \approx mv = 10^{-30} \text{ kg} \times 2 \times 10^6 \text{ m/s} = 2 \times 10^{-24} \text{ kg} \times \text{m/s}.$$

We now substitute this value for p and the value for Planck's constant, Eq. (IV–5), into Eq. (IV–24):

$$\lambda = \frac{h}{p} = \frac{6.6 \times 10^{-34} \text{ j} \times \text{s}}{2 \times 10^{-24} \text{ kg} \times \text{m/s}} = 3 \times 10^{-10} \text{ meters.}$$

This length is roughly the diameter of an atom. Thus the wavelike character of the electron causes the electron to be "spread out" over the entire atom.

If you have trouble figuring the units in this problem, just remember that if you stick to kilograms, meters, and joules, everything will come out right.

6. The Heisenberg Uncertainty Principle

We are now ready to combine all of our previous results to obtain Heisenberg's uncertainty principle, Eq. (IV–28). As we have done so many times before in this chapter, consider a stream of particles passing through a hole in a shade. The particles could be photons or electrons or any other kind of matter or energy. Suppose that the particles are all traveling in the horizontal direction before meeting the shade and that they all have the same momentum p. Before the particles pass through the hole, their momentum is completely in the horizontal direction. After the particles pass through the hole, their momentum will be partly in the horizontal direction and partly in the vertical and left-right directions, since the particles will spread out as a result of the wavelike part of their personality. The situation is shown in Fig. IV–14, where the spreading angle θ is shown. Note that this figure shows the spreading only in the vertical direction, and does not include the spreading in the left-right direction. The particles spread out by the same angle θ in the left-right direction. (See Fig. IV–13a.) For the moment, we will consider the spreading only in the vertical direction.

If we combine Eqs. (IV–17), $\tan \theta \approx \lambda/2a$, and (IV–20), $\tan \theta = p_y/p_h$, noting that the same angle θ appears on the left-hand side of both equations, we obtain the relation

$$\frac{\lambda}{2a} \approx \frac{p_y}{p_h}. \qquad\qquad \text{(IV–25a)}$$

Equation (IV–25a) relates the wavelength λ of the incoming particles to the width a of the hole and to their vertical and horizontal momenta p_y and p_h. (Here p_y is the vertical momentum of those emerging particles that are deflected *the most* in passing through the hole. There will be some particles that are deflected much less.) Solving Eq. (IV–25a) for p_y, we get

$$p_y \approx \frac{p_h \lambda}{2a}. \qquad\qquad \text{(IV–25b)}$$

Initially, all the particles were traveling in the horizontal direction, so that they all had zero vertical momentum, $p_y = 0$. After the particles pass

Figure IV–14: Spreading of particles after emerging through a hole, showing the spreading angle θ.

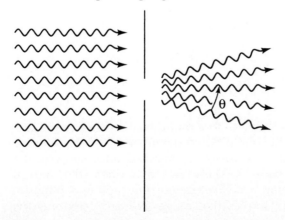

through the hole, they gain a range of vertical momenta, extending from 0 to $p_h\lambda/2a$, the maximum vertical momentum, given by Eq. (IV–25b). It is impossible to say what any given particle's vertical momentum will be within this range. If we denote this range of momenta by δp_y, then

$$\delta p_y \approx \frac{p_h\lambda}{2a}. \tag{IV–26a}$$

Equation (IV–26a) can be simplified. Recall that wavelength is related to momentum through de Broglie's relation, Eq. (IV–24), $\lambda = h/p$, so that we can eliminate the λ from Eq. (IV–26a). For the p in Eq. (IV–24) we can substitute p_h, since the momentum is initially all in the horizontal direction. Substituting $\lambda = h/p_h$ into Eq. (IV–26a), we then obtain

$$\delta p_y \approx \frac{h}{2a}. \tag{IV–26b}$$

Equation (IV–26b) expresses an uncertainty in the vertical component of momentum of the particles emerging from the hole. Before passing through the hole, each particle had a certain momentum, $p_y = 0$ and $p_h = p$. After passing through the hole, each particle can have any vertical momentum from 0 to δp_y, that is, its vertical momentum is uncertain and δp_y is the amount of uncertainty.

Each particle emerging from the hole also has an uncertain vertical height. A particle could start anywhere from the bottom of the hole to the top of the hole. This range of uncertainty in vertical height is just the width of the hole, a. If we denote the uncertainty in vertical height by δy, then we have just concluded that

$$\delta y = a. \tag{IV–27}$$

Multiplying Eq. (IV–26b) by a and substituting $a = \delta y$ from Eq. (IV–27), we finally obtain

$$\delta p_y \times \delta y = \frac{h}{2}, \tag{IV–28a}$$

where, for clarity, we have explicitly put in the multiplication sign on the left-hand side. Equation (IV–28a) is the quantitative statement of the Heisenberg uncertainty principle. The equation expresses a combined uncertainty in the position and momentum of a particle. These two quantities—the vertical height and the vertical momentum—can be known only within a range of uncertainty. Equation (IV–28a) quantifies that range. We can determine the vertical height very accurately, making δy very small, but then the uncertainty in vertical momentum becomes large because Eq. (IV–28a) says that

$$\delta p_y = \frac{h}{2\delta y}. \qquad \text{(IV–28b)}$$

Or, we can determine the vertical momentum very accurately and make δp_y very small, but then the uncertainty in vertical position will get large because Eq. (IV–28a) says that

$$\delta y = \frac{h}{2\delta p_y}. \qquad \text{(IV–28c)}$$

Now recall that we have ignored the spreading in the left-right direction. But this is easy. Since the hole in the shade is circular, the situation in the left-right direction is identical to that in the vertical direction, including the initial uncertainty in the left-right position of the particle, $\delta z = a$, and the uncertainty in left-right momentum, δp_z. Thus, by analogy to Eq. (IV–28a), we obtain

$$\delta p_z \times \delta z = \frac{h}{2}. \qquad \text{(IV–28d)}$$

Equation (IV–28d) shows how the uncertainty in left-right position and uncertainty in left-right momentum are related, just as Eq. (IV–28a) shows how the uncertainties in vertical position and vertical momentum are related.

In short, if we want to predict accurately the future trajectory of a particle after it emerges from a hole, then we have troubles. As you remember from section F–1, predictions of trajectories require knowledge of *both* initial position and initial speed. Since speed and momentum are intimately connected, initial momentum can substitute for initial speed in making predictions. Before quantum theory, physicists believed that both the initial position and the initial momentum of a particle could be determined as accurately as we want, with no uncertainty, making $\delta y = 0$, $\delta z = 0$, and $\delta p_y = 0$, $\delta p_z = 0$. We could arrange for an infinitely small hole *and* for all the particles to emerge from the hole with zero vertical and left-right momenta. But the Heisenberg uncertainty principle says we can't do that. There is an unavoidable tradeoff between knowledge of position and knowledge of momentum.

The Heisenberg uncertainty principle is a necessary consequence of the wave-particle duality of nature. The amount of the uncertainty is fixed by Planck's constant. If Planck's constant were zero, then the wavelength of any particle would be zero, as seen by Eq. (IV–24), and the wavelike properties of a particle would disappear. Equation (IV–28a) would reduce to $\delta y \times \delta p_y = 0$; Eq. (IV–28d) would read $\delta z \times \delta p_z = 0$. In other words, uncertainties in both y

and p_y could be reduced to zero, and the same for z and p_z. However, Planck's constant is not zero. We cannot completely predict the future trajectory of a particle, or anything else. We can only assign probabilities to various trajectories, based on the range of uncertainty in the initial conditions of the particle. The Heisenberg uncertainty principle means that we must forever live with probabilities, not certainties.

Problem IV–6: Molecules Through a Hole

Suppose that a stream of oxygen molecules moves toward a hole of diameter 10^{-8} meters, about 10 times the diameter of a molecule. The molecules are all moving in the horizontal direction before passing through the hole. Each molecule has a mass of 2×10^{-26} kilograms and an initial speed of 300 meters per second (the speed of sound in air). (a) What is the initial momentum of a molecule? (b) What is the uncertainty in momentum of a molecule after it emerges from the hole?

Solution: (a) The molecules are moving at a speed much smaller than the speed of light, and so we can use the approximation of Eq. (IV–22):

$$p \approx mv = (2 \times 10^{-26} \text{ kg})(300 \text{ m/s}) = 6 \times 10^{-24} \text{ kg} \times \text{m/s}.$$

(b) The uncertainty in vertical height, δy, is just the width of the hole, $\delta y = 10^{-8}$ meters. Substituting this into Eq. (IV–28b), we obtain

$$\delta p_y = \frac{h}{2\delta y} = \frac{6.6 \times 10^{-34} \text{ j} \times \text{s}}{2 \times 10^{-8} \text{ m}} = 3.3 \times 10^{-26} \text{ kg} \times \text{m/s}.$$

Note that the fractional uncertainty in momentum is

$$\frac{\delta p_y}{p} = \frac{3.3 \times 10^{-26} \text{ kg} \times \text{m/s}}{6 \times 10^{-24} \text{ kg} \times \text{m/s}} = 0.006.$$

Problem IV–7: Predictability in the Quantum World

Suppose that a lion has just charged you and you fire an electron gun at him. Your life depends on accurately predicting the trajectory of the electron bullets. Equation (IV–8) gives the vertical height y_f of a bullet after an elapsed time Δt in terms of the initial height y_i and the initial upward speed v_i:

$$y_f = y_i + v_i \Delta t - \frac{1}{2} g (\Delta t)^2.$$

If the bullets are traveling at a speed small compared with the speed of light, then the initial vertical speed of a bullet is approximately equal to its initial vertical momentum p_{yi} divided by its mass m, [Eq. (IV–22)]:

$$v_i = \frac{p_{yi}}{m}.$$

Substituting this result into the previous equation, we obtain

$$y_f = y_i + \frac{p_{yi}\,\Delta t}{m} - \frac{1}{2}g\,(\Delta t)^2. \qquad\qquad \text{(IV–29)}$$

Now, if there is an uncertainty in the initial vertical height, δy_i, and an uncertainty in the initial vertical momentum, δp_{yi}, the uncertainty in final height of the bullet after an elapsed time Δt is

$$\delta y_f = \delta y_i + \frac{\delta p_{yi}\,\Delta t}{m}. \qquad\qquad \text{(IV–30)}$$

Equation (IV–30) follows from Eq. (IV–29). (Notice that the gravitational term, $g\,(\Delta t)^2/2$, has no uncertainty and thus does not add to the uncertainty in y_f.) Either an uncertainty in y_i or an uncertainty in p_{yi} will cause an uncertainty in the vertical position of each bullet at the time it reaches the lion. The situation is illustrated in Fig. IV–15. Each bullet can originate anywhere within a region of vertical height δy_i, with a range of directions corresponding to the uncertainty in vertical momentum, δp_{yi}. All we can say is that the final height of the bullet, at the time it reaches the distance of the lion, will lie somewhere within the region of impact shown in Fig. IV–15b. Notice that again for simplicity Fig. IV–15b has left out the spreads and uncertainties in the left-right direction.

(a) Suppose that the bullets reach the lion after 0.1 second. Each bullet, being an electron, has a mass of 10^{-30} kilogram. Given the Heisenberg uncertainty principle, what is the *minimum possible* value of δy_f? Would this uncertainty in vertical height of the bullets be small enough to ensure killing the lion?

(b) Repeat the above for bullets of mass 0.01 kilogram, which is much closer to the mass of a standard rifle bullet.

This is a long and difficult problem, but it illustrates well the application of the Heisenberg uncertainty principle. If you have trouble working the problem, study the solution.

Solution: Using the Heisenberg uncertainty principle in the form of Eq. (IV–28b), we can express the uncertainty in initial vertical momentum δp_{yi} in terms of the uncertainty in initial vertical height:

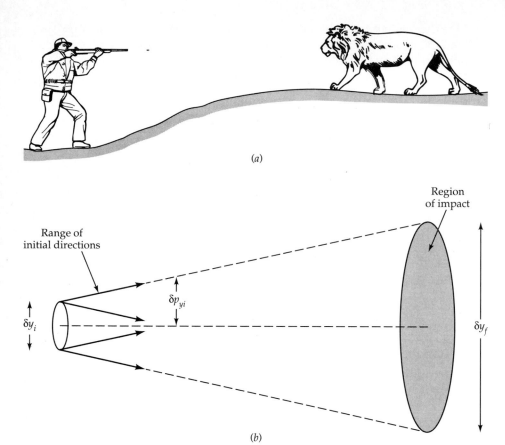

Figure IV–15: (*a*) A person attempting to kill a charging lion with a gun that fires electron bullets. (*b*) Analysis of the uncertainty in position of the bullets when they reach the lion, in terms of the initial uncertainties in position and momentum of the bullets.

$$\delta p_{yi} = \frac{h}{2\delta y_i},$$

where we have just used δp_{yi} for δp_y and so on. Substituting this relation into Eq. (IV–30), we find

$$\delta y_f = \delta y_i + \frac{h \, \Delta t}{2m \, \delta y_i}. \qquad \text{(IV–31)}$$

Equation (IV–31) shows that δy_f becomes large if δy_i is either very small or very large. If we make δy_i very small, so that the first term on the right-hand side is small, then the second term, associated with the uncertainty in initial vertical momentum, becomes very big. Physically, we are locating the initial position of the bullet very accurately, but at the expense of a very inaccurate initial direction of motion. In other words, we know where the gun is, but our aim is wild. On the other hand, if we want our initial direction of motion to be very accurate and so make δy_i large, so that the second term is small, then the first term, associated with the uncertainty in initial position of the bullet, becomes very large. In summary, if δy_i is either very large or very small, then we are stuck with a large value of δy_f, the uncertainty in final height of the bullet. Evidently, there is some intermediate value of δy_i that makes δy_f as small as possible. That's what we want to find.

A trick will help us find the minimum value of δy_f. It is possible to rewrite Eq. (IV–31) in the form

$$\delta y_f = 2\left(\frac{h\,\Delta t}{2m}\right)^{1/2} + \frac{1}{\delta y_i}\left[\delta y_i - \left(\frac{h\,\Delta t}{2m}\right)^{1/2}\right]^2. \qquad \text{(IV–32)}$$

To see that Eq. (IV–32) is equivalent to Eq. (IV–31), first carry out the square in the second term on the right-hand side:

$$\left[\delta y_i - \left(\frac{h\,\Delta t}{2m}\right)^{1/2}\right]^2 = \left[\delta y_i - \left(\frac{h\,\Delta t}{2m}\right)^{1/2}\right] \times \left[\delta y_i - \left(\frac{h\,\Delta t}{2m}\right)^{1/2}\right]$$

$$= (\delta y_i)^2 - 2\delta y_i\left(\frac{h\,\Delta t}{2m}\right)^{1/2} + \frac{h\,\Delta t}{2m}.$$

Substituting this result into Eq. (IV–32), where we have to divide by δy_i, we get

$$\delta y_f = 2\left(\frac{h\,\Delta t}{2m}\right)^{1/2} + \delta y_i - 2\left(\frac{h\,\Delta t}{2m}\right)^{1/2} + \frac{h\,\Delta t}{2m\,\delta y_i}.$$

Now, notice that in this last equation, the first and third terms on the right-hand side cancel, leaving

$$\delta y_f = \delta y_i + \frac{h\,\Delta t}{2m\,\delta y_i}.$$

But this is identical to Eq. (IV–31), proving that Eq. (IV–32) is indeed equivalent to Eq. (IV–31).

Now notice that the squared term on the right-hand side of Eq. (IV–32) can never be negative. (Any number multiplied by itself is positive.) The *smallest* that term can be is zero, which occurs when

$$\delta y_i = \left(\frac{h \, \Delta t}{2m}\right)^{1/2}.$$

When δy_i has this value, then δy_f is the smallest it can be,

$$\delta y_f = 2 \left(\frac{h \, \Delta t}{2m}\right)^{1/2}. \qquad \text{(IV–33)}$$

Equation (IV–33) gives the *minimum possible* value of the uncertainty in final height of the bullet consistent with the Heisenberg uncertainty principle. Now, we are ready to substitute in numbers.

(a) Substituting the given values of Δt, m, and h into Eq. (IV–33), we get

$$\delta y_f = 2 \left(\frac{(6.6 \times 10^{-34} \text{ j} \times \text{s})(0.1 \text{ s})}{2 \times 10^{-30} \text{ kg}}\right)^{1/2} = 1.1 \times 10^{-2} \text{ m} = 1.1 \text{ cm}.$$

(This translates to about half an inch.) If we aimed our gun right at the center of the lion's heart, which is about 10 centimeters in width, an error of 1.1 centimeter would still place the bullet in the heart. So the inaccuracy in the bullet's trajectory due to the Heisenberg uncertainty principle would not prevent us from killing the charging beast and saving ourselves! However, we aren't safe by much. If the lion's heart were a few times smaller, then we would likely miss it, no matter how accurately we tried to control the path of our bullet.

(b) Notice that Eq. (IV–33) says that the minimum δy_f varies inversely with the square root of the mass of the bullet. Thus, a bullet with four times the mass should produce $1/\sqrt{4} = 1/2$ the value of δy_f. A bullet of mass $m = 0.01$ kilogram is $10^{-2}/10^{-30} = 10^{28}$ times more massive than a bullet of mass 10^{-30} kilogram and thus leads to $1/\sqrt{10^{28}} = 10^{-14}$ the value of δy_f. Thus, for this more massive bullet,

$$\delta y_f = 1.1 \times 10^{-2} \text{ meters} \times 10^{-14} = 1.1 \times 10^{-16} \text{ meters}.$$

This is an incredibly tiny uncertainty, smaller than the width of an atom, and gives far more accuracy than needed to be sure of killing the lion.

We can see that the Heisenberg uncertainty principle has a much bigger effect for smaller masses. For the masses of everyday experience, the principle is almost irrelevant. However, for small masses, such as the electron and other subatomic particles, the uncertainty dictated by the principle can be significant. That is why it is often said that quantum mechanical effects are important in the subatomic world but not in the macroscopic world.

7. Quantum Mechanics

Ideas about the wave-particle duality of nature, the de Broglie relation, and the Heisenberg uncertainty principle were all incorporated into a detailed theory called quantum mechanics, formulated in the late 1920s. The chief architects of quantum mechanics were Heisenberg and the Austrian physicist Erwin Schrödinger (1887–1961).

   Quantum mechanics is a probabilistic theory of nature. The equations of the theory describe the average behaviors of a large collection of identical particles but do not describe the behavior of individual particles. Despite this restriction, quantum mechanics has been extremely successful at predicting the results of laboratory experiments involving atomic processes. As we have shown, the effects of quantum phenomena are important at the atomic and subatomic level but not so important in the world of ordinary experience. Quantum mechanics has successfully described such diverse phenomena as the light emitted by atoms, the workings of transistors, the chemistry of water, and the existence of new subatomic particles, such as quarks. Modern physics is built upon quantum mechanics.

G. DETERMINISM, CAUSALITY, AND CHOICE IN THE QUANTUM WORLD

The understanding of quantum physics has called into question our ability to make accurate predictions of the future and, more fundamentally, whether the future is determined by the present. Scientists, philosophers, and others have strongly reacted to this new view of the world. We will consider some of those reactions here.

1. Einstein's View of Quantum Mechanics

Einstein was very impressed by the successes of quantum mechanics, but he had deep reservations about the theory. In opposition to the demands of quantum mechanics, Einstein did not believe that nature is intrinsically uncertain. In a letter to the physicist Max Born, written on December 4, 1926, Einstein wrote

   Quantum mechanics is very impressive. But an inner voice tells me that it is not yet the real thing. The theory produces a great deal but hardly brings us closer to the secret of the Old One. I am at all events convinced that *He* does not play dice. [*The Born-Einstein Letters* (Walker: New York, 1971), p. 90]

Einstein considered quantum mechanics to be a good approximation to nature but an incomplete theory. When quantum mechanics became complete, according to Einstein, it would deal with certainties. An improved theory would allow the certain prediction of the trajectory of a single electron and the certain prediction of the precise moment when a single atom would disintegrate.

Discussion Questions IV–5

What did Einstein mean when he said that the Old One does not play dice? In what way is nature, as pictured by quantum physics, a dice game? Was Einstein's objection to quantum mechanics physical or philosophical? From the above quote, do you think that Einstein believed in the possibility of a perfect theory of nature?

---

2. Causality

In section F–1, we discussed the belief that the future is completely determined by the present. This idea was given a rigorous scientific basis by the work of such physicists as Newton and Laplace in the seventeenth and eighteenth centuries and has been debated by philosophers for centuries. Determinism is closely connected to the concept of causality, which states that every event has a prior cause.

The following passage by the Czech-American philosopher Milic Capek, from a chapter titled "The End of the Laplacian Illusion," is indicative of the reaction of some philosophers to the Heisenberg uncertainty principle:

The principle of uncertainty, formulated by Heisenberg, forbids a simultaneous knowledge of the position and velocity of any elementary particle. . . . We have only to remember that in the classical Laplacian model of the universe "the world at a given instant" was definable as a huge instantaneous configuration of elementary particles, each possessing besides its definite mass also a sharply defined position and velocity; a "state of the world" thus defined contained all past and future configurations and velocities because any event in world history was in principle deducible from any sharply defined cosmic state. According to the principle of uncertainty, both the concept of precise position and that of sharply defined velocity lose their meanings; consequently, the concept of the "state of the world at an instant" loses its definiteness too. [Milic

Capek, *The Philosophical Impact of Contemporary Physics* (D. Van Nostrand: Princeton, 1961) pp. 289-290]

Discussion Questions IV–6

Is there a difference between (a) our inability to define the world at any given instant and (b) the world's instrinsic lack of definiteness and precision at any instant? If so, which of the two does the Heisenberg uncertainty principle refer to? Which of the two does causality depend on? Could the world be causal but unpredictable?

Discussion Questions IV–7

Why would a philosopher be concerned with determinism and causality? Do you think that philosophers should incorporate physics into their thinking?

---

3. Determinism, Free Will, and Ethics

A longstanding debate among philosophers and theologians is whether human beings are able to act by choice, out of their own free will, or whether they are only elaborate machines, doomed to obey forces and blueprints beyond their control. This debate spills over into ethics. In the view of many philosophers and theologians, the question of ethical or unethical behavior loses its meaning if human beings do not have the ability to freely choose between right and wrong.

The Heisenberg uncertainty principle, by eliminating rigid determinism, would seem to allow more choice. The question is considered here by the German philosopher Ernst Cassirer (1874–1945), in a book written less than a decade after the uncertainty principle was formulated.

> One of the essential tasks of philosophical ethics consists of showing . . . why freedom does not need to be upheld against physical causality but instead maintains and asserts itself on its own grounds. "Persistence" is not only a physical but also and at the same time an ethical category, although in an entirely different sense. For all truly ethical actions must spring from the unity and persistence of a definite ethical character. This in itself shows us that it would be fatal for ethics to tie itself to and, as it were, fling itself into the arms of a limitless indeterminism. From such a standpoint we would have to evaluate an action more highly the more it bears the earmark of the arbitrary, the unforseen, the unpredictable. Yet true ethical judgement runs in exactly the opposite direction. . . .

Ethical character is distinguished by the fact that it is not completely determined from the outside, that in its decisions it is not thrown hither and thither by the changing conditions of the moment but remains itself and persists in itself. . . . Thus it becomes clear from this side as well that a possible change of the physical "causality concept" cannot directly touch ethics. For however physics may change its internal structure, by abandoning, for example, the concept of the simple mass point [particle] or the possibility of strict predictions, the opposition in principle between the physical and ethical world, between the realm of nature and the realm of ethics, cannot be bridged. [Ernst Cassirer, *Determinism and Indeterminism in Modern Physics* (Goteborgs Hogskolas Arskrift: Sweden, 1936), trans. O. T. Benfey (Yale University Press: New Haven, 1956), pp. 204–205]

Discussion Questions IV–8

In Cassirer's view, what determines an ethical judgment? Do you agree? What does Cassirer mean when he says that ethical judgment "is not completely determined from the outside"? What is the "outside" that Cassirer refers to, and what would be the corresponding "inside"? In Cassirer's view, does the indeterminacy of quantum physics apply to the "outside" or the "inside"? Does Cassirer give a convincing argument as to why the realm of nature and the realm of ethics cannot be bridged? What is your view?

The Heisenberg uncertainty principle sent shocks through the public almost immediately. The following passages are from an article titled "The Uncertainty Principle and Human Behavior," which appeared in *Harper's* magazine in 1935 and which illustrates the humanistic impact of major scientific ideas.

Seventy-five years ago the imagination of the world was kindled and the thought of a generation stirred by Darwin's contribution to theoretical biology. Evolution became the keyword to discussion and investigation even in the most remotely relevant of human activities. Today the center of a possibly similar intellectual disturbance is located in theoretical physics. The ideas recently emanating from this source have captured the popular imagination and have begun to permeate the tissue and structure of our thought in surprisingly varied ways. (p. 237)

A physical object is something which stays put unless something else starts it going, whereas an animal is something which under any circumstances does what it pleases. Isn't a man a living creature and cannot he do pretty much as he pleases? . . . Heretofore, even if man entered somewhat into the matter, at least a given cause under a given set of conditions always produced the same effect. But along came Heisenberg with his principle of indeterminancy and apparently destroyed the pure and inevitable relations of cause and effect. In the body of physics itself this new principle produced the happiest of results. . . . But outside of physics, and in the minds of even some physicists, the devastation produced has been pitiful. (p. 238)

Chemical changes are electron changes; and there are probably so few electron changes in these very small units of biological structure that they come under the uncertainty principle. Therefore, the very basis of biological behavior is unpredictable. Usually the next step in the argument is that this uncertainty leaves room for and indeed demands the existence of free will in human behavior. (p. 241)

Why then do we not lay the claims of biological mechanism before the public? Largely, I suppose, because we think it irrelevant. What does it matter whether we publicly uphold mechanism or vitalism [internal life force and choice]? . . . The pursuit of science has no bearing on this mechanism-vitalism discussion; it does not deal with free will and determinism. The method of science, as always, is independent of these ideas. Measurements, verification, ideas, computations—these are the stuff of science (p. 242)

On the analytical level of the laboratory, the mind sees things as determined. (p. 246) . . . But in human behavior, in our daily lives, the decision is not simple. To his own mind, the behavior of a man seems to be free and of his own choosing, and all the accumulated moralities of the world exhort him to choose the good and to act righteously on the assumption that he is capable of free choice and action. (p. 247)

What then are we to do as human beings in a social order? Clearly we have to be biological units *and* biological investigators; we need free will *and* determinism. This sounds contradictory and inconsistent. But so is the hard and solid table which turns out to be mostly empty space. Perhaps it will seem less contradictory if instead of free will and determinism I say instinct and reason; for that is precisely what it comes down to. (p. 249) [Selig Hecht, " The Uncertainty Principle and Human Behavior," *Harper's*, January 1935]

Discussion Questions IV–9

Do you think Hecht believes in the validity of the Heisenberg uncertainty principle? If so, what is the realm of its validity in Hecht's view? Does he say whether the principle allows freedom in human actions? According to Hecht, is the Heisenberg uncertainty principle relevant to the issue of free will versus determinism in human actions? If so, how is it relevant? If not, why does Hecht bring up the principle in his discussion? Compare his conclusions with those of Cassirer in the previous excerpt. If issues of human behavior are much more complex than the behavior of atoms and electrons, why are Cassirer and Hecht concerned at all about the developments in quantum physics? Do you think that humanists should be concerned with new ideas in science?

Readings

George Berkeley, *Three Dialogues Between Hylas and Philonous* (1713) (Bobbs-Merrill: New York, 1954).

Niels Bohr, "Discussion with Einstein on Epistemological Problems in Atomic Physics," in *Albert Einstein: Philosoper Scientist*, ed. P. Schilpp (Tudor: New York, 1949).

Ernst Cassirer, *Determinism and Indeterminism in Modern Physics*, trans. O. T. Benfey (Yale University Press: New Haven, 1956).

Louis de Broglie, "The Undulating Aspects of the Electron," Nobel Prize Address 1929, in *The World of the Atom*, ed. H. A. Boorse and L. Motz (Basic Books: New York, 1966).

Selig Hecht, "The Uncertainty Principle and Human Behavior," *Harper's* January 1935, p. 237.

Werner Heisenberg, *Physics and Philosophy* (Harper: New York, 1958).

Hans Reichenbach, *Philosophical Foundations of Quantum Mechanics* (University of California Press: Berkeley, 1982), reprint of 1950 edition.

Erwin Schrödinger, *Science, Theory, and Man* (George Allen and Unwin: London, 1935).

Laboratory

A. PURPOSE

The purpose of this laboratory is to find out what parameters determine the period of a pendulum, to quantify the relationship between the period and those parameters, and to use the result to make predictions.

B. EQUIPMENT

The lab takes about 2 hours and requires several sets of pendulums with adjustable lengths and weights, meter sticks, and stopwatches. A convenient pendulum support can be made out of a long metal rod, a short metal rod, and two "C clamps." One end of the long rod is clamped to the edge of a table, firmly holding the rod in a vertical position. The short rod, from which hangs the pendulum string, is held horizontally from the top of the long rod with the second clamp. The pendulum string is attached to the end of the short rod, either by a knot or by some squeeze device. However it is attached, the pendulum string should be easily adjustable in length, say from 6 inches to 6 feet. A weight holder is attached to the bottom of the pendulum string, and weights of various amounts are inserted there.

C. PROCEDURE

Divide the class into several groups. Each group is given a pendulum with string, several weights, a meter stick, and a stopwatch. First, the *period* of a pendulum must be defined. Set one of the pendulums swinging and measure with a stopwatch the length of time it takes the pendulum to make one complete swing, returning to the point where it was released. This time for one complete swing, back and forth, is called the period. For accuracy, it is best to determine the period by measuring the time it takes the pendulum to complete five cycles, and then dividing the result by 5.

Do all pendulums have the same period? At this point, and *before any experiments*, each group of students should privately discuss among themselves what factors determine the period of a pendulum. Be as quantitative as possible. Each group must come to a consensus and write its predictions on a piece of paper. The predictions from each group are then collected by the instructor.

Now, the experiments begin. Each group should do whatever experiments it thinks relevant to testing its predictions. The experiments should be quantitative, using the stopwatches, weights, and meter sticks to measure the period under various conditions, and the results should be recorded. This portion of the lab will take about an hour.

After the data have been taken, the instructor should collect and combine the data from all of the groups and graph the results on the blackboard. Compare the actual results with the predictions and discuss. Ideally, the period should be measured over a large enough range of length, weight, swing amplitude, and other parameters so that definite and quantitative laws are easily seen. From the data, derive an empirical relationship between the period and other parameters.

Now use this empirical relationship to predict the period of a pendulum of a length not tested by any of the class groups. Ideally, this should be a length either shorter than the shortest length already tested or longer than the longest length tested. Construct such a pendulum, measure its period, and compare to the prediction. Discuss the predictive power of universal laws of nature and the meaning of such laws.

APPENDIX A

A Review of Some Basic Mathematics

1. NUMBERS AND SCIENTIFIC NOTATION

Because science frequently involves very big numbers or very small numbers, a shorthand notation has been developed, as illustrated in the following examples:

$$10^4 = 10,000$$

$$10^6 = 1,000,000.$$

The superscript number to the right of the 10 tells how many zeros follow the 1. This notation can then be used to express other numbers in the following way:

$$2 \times 10^4 = 2 \times 10,000 = 20,000$$

$$3.48 \times 10^6 = 3.48 \times 1,000,000 = 3,480,000,$$

and so on. The superscript number to the right of the 10 tells how many places to move the decimal point to the right in the preceding number.

The number 1 can be represented by

$$1 = 10^0,$$

since there are 0 zeros following the 1. For numbers smaller than 1, we use *negative* superscripts:

$$10^{-1} = 0.1$$

$$10^{-3} = 0.001$$

$$10^{-8} = 0.00000001,$$

and so on. When the superscript is negative, it means to move the decimal point to the *left* that many places in the preceding number. For example,

$$2.72 \times 10^{-5} = 0.0000272.$$

Positive superscripts produce large numbers and negative superscripts produce small numbers. The width of a hydrogen atom is about 1.5×10^{-8} centimeters, and the distance to the sun is about 1.5×10^{13} centimeters.

When the number 1 multiplies the 10, the 1 is usually omitted. For example, 1×10^6 is written as 10^6, which equals 1,000,000.

2. EXPONENTS

a. Definition of Exponents

When a number is multiplied by itself many times, the result is abbreviated with an exponent, as shown in the following examples:

$$3 \times 3 \times 3 \times 3 = 3^4$$

$$7 \times 7 \times 7 \times 7 \times 7 = 7^5.$$

In the number 3^4, the 3 is called the base and the 4 is called the exponent. In the number 7^5, the base is 7 and the exponent is 5. The exponent tells how many times the base is multiplied by itself. We call the number 3^4 "three raised to the power 4," and so forth.

If the exponent is negative, it means to take the reciprocal. For example,

$$3^{-4} = \frac{1}{3^4} = \frac{1}{3 \times 3 \times 3 \times 3} = \frac{1}{81}$$

$$7^{-5} = \frac{1}{7^5} = \frac{1}{7 \times 7 \times 7 \times 7 \times 7} = \frac{1}{16,807}.$$

In scientific notation, the 10 is the base and the superscript is the exponent. We can think of the number 1.26×10^3, for example, as $1.26 \times 10 \times 10 \times 10 = 1.26 \times 1000 = 1260$.

b. Manipulating Numbers with Exponents

When two numbers with the same base are multiplied, their exponents are added:

$$3^4 \times 3^2 = (3 \times 3 \times 3 \times 3) \times (3 \times 3) = 3 \times 3 \times 3 \times 3 \times 3 \times 3 = 3^6 = 3^{4+2}$$

$$7^2 \times 7^{-4} = (7 \times 7) \times \left(\frac{1}{7 \times 7 \times 7 \times 7}\right) = \frac{1}{7 \times 7} = 7^{-2} = 7^{2-4}.$$

When a number with an exponent is raised to a power, the two exponents are multiplied:

$$(3^4)^3 = (3 \times 3 \times 3 \times 3) \times (3 \times 3 \times 3 \times 3) \times (3 \times 3 \times 3 \times 3) = 3^{12} = 3^{4 \times 3}.$$

c. Fractional Exponents

Square roots are familiar uses of fractional exponents. For example, the square root of 16, usually written as $\sqrt{16}$, is the number that equals 16 when multiplied by itself. There are two such numbers: $\sqrt{16} = 4$ or -4, because $4 \times 4 = 16$ and $-4 \times -4 = 16$. The square root of 16 can also be written as $16^{0.5}$, in which 16 is the base and 0.5 is a fractional exponent.

By analogy with the square root, $8^{1/3}$ is the number that equals 8 when multiplied by itself *three times*, or raised to the power 3. Thus $8^{1/3} = 2$ because $2 \times 2 \times 2 = 2^3 = 8$. This number can also be written as $8^{0.3333}$, in which 0.3333 is a fractional exponent.

Numbers resulting from fractional exponents do not have to be whole numbers. For example,

$$9^{1/3} = 2.0801,$$

accurate to the first five digits. The correctness of this result can be verified by raising 2.0801 to the power 3 and obtaining 9, to the accuracy of the first five digits. In general, numbers with fractional exponents have to be calculated with an electronic calculator. We won't have to worry much about calculating such numbers in this course.

Fractional exponents can be larger than 1:

$$9^{1.3333} = 18.72.$$

Using the rule for multiplying two numbers with the same base, $9^{1.3333} = 9 \times 9^{0.3333}$. In this way, a fractional exponent larger than 1 can always be recast into a whole number exponent and a fractional exponent less than 1.

Negative fractional exponents are treated just like negative whole number exponents. For example,

$$9^{-1/3} = \frac{1}{9^{1/3}}.$$

3. EQUATIONS

a. Definitions and Rules for Manipulating Equations

Equations are simply statements that two things are equal. For example,

$$2 \times 3 + 7 = 4 + 3 \times 3$$

is an equation. Each number that is added to another number in an equation is called a *term*. In the above example, 2×3 is the first term on the left side of the equation and 7 is the second term. Thus, there are two terms on each side of the equation. Note in the above example that the multiplications are carried out first, and then the additions. Thus, on the left-hand side, the 2 is multiplied by 3 to give 6, and then the 6 is added to the 7.

Now we come to the rules for manipulating equations. If two equal things are multiplied by the same number, the results are also equal. Consider, for example, the equation

$$3 \times 7 = 21.$$

If both sides of this equation are multiplied by 2, we get

$$2 \times 3 \times 7 = 2 \times 21,$$

which is also a correct equation. Thus, starting with any correct equation, multiplying both sides of it by the same number produces another correct equation. Likewise, both sides of an equation can be divided by the same number. Starting with the equation

$$6 \times 7 = 3 \times 14,$$

we can divide both sides by 3:

$$\frac{6 \times 7}{3} = \frac{3 \times 14}{3},$$

which is also a correct equation. Likewise the same number can be added or subtracted from both sides of an equation. For example, starting with

$$8 \times 7 = 56,$$

we can subtract 3 from both sides of the equation, obtaining the correct equation

$$8 \times 7 - 3 = 56 - 3.$$

Multiplication, division, addition, and subtraction are examples of operations. *Any* identical operation may be performed on both sides of an equation. "Equals done to equals produce equals." For example, both sides of an equation may be raised to the same power. Consider the equation

$$3 \times 4 = 12.$$

We can raise both sides to the power 0.2, obtaining

$$(3 \times 4)^{0.2} = (12)^{0.2},$$

which is also a correct equation.

b. Symbolic Notation

It is often inconvenient to use numbers in every result or equation. For example, suppose we want to state the rule for multiplying two numbers with the same base. We previously gave as an illustration $3^2 \times 3^4 = 3^6$ and then said that the exponents are added in such a situation. A general equation expressing this rule is

$$y^a \times y^b = y^{a+b},$$

where y, a, and b can be *any* numbers whatsoever. By using the symbols y, a, and b, we can express a result that describes all numbers and not simply the ones in our particular examples. Symbolic notation is a powerful tool for expressing general results.

Symbolic notation is also useful for relating physical quantities in a general way. In the formula for kinetic energy, for example, we write

$$E_K = \tfrac{1}{2}\, mv^2,$$

where E_K stands for kinetic energy, m stands for mass, and v for speed. Regardless of the actual numbers for the mass and speed of any particular object, this equation says that the kinetic energy of that object is equal to one-half its mass multiplied by the square of its speed. Again, the equation expresses a generality, allowed by the use of symbolic notation.

c. Solving Equations with Unknowns

Sometimes there are terms in an equation that we do not know in advance. The unknown terms are expressed as symbols, and the equation is solved for the unknown, as in the example

$$x + 7 = 12.$$

Here, x stands for the unknown. To solve this equation for x, subtract 7 from both sides, obtaining the solution $x = 5$. A more complicated equation with an unknown is

$$\frac{x^4}{73} + 18 = 87.$$

To solve this equation for x, first subtract 18 from both sides and then multiply both sides by 73, obtaining

$$x^4 = (87 - 18) \times 73 = 5037.$$

Now raise each side to the power 1/4:

$$x^{4 \times 1/4} = x^1 = x = (5037)^{1/4} = 8.42.$$

Thus $x = 8.42$ is the solution of this equation.

d. The Multiplication Convention

Symbols are multiplied so often in equations that it is a nuisance to keep repeating the multiplication sign $\times$. Therefore, this sign is often omitted. Whenever two quantities, or symbols, appear side by side, they are meant to be multiplied. For example, ab means $a \times b$. For the remainder of this appendix, we will use this convention.

e. The Use of Parentheses

A set of parentheses around several terms or numbers means that any operation, such as multiplication or squaring, outside the parentheses acts on all of the terms inside the parentheses. For example,

$$a(b + c + d) = ab + ac + ad.$$

Equivalently, in the above example, the terms a, b, and c are added together first and then the sum is multiplied by a.

As another example,

$$(a + b)^2 = (a + b) \times (a + b) = a^2 + ab + ba + b^2 = a^2 + 2ab + b^2.$$

4. THE QUADRATIC FORMULA

Equations with several terms, each containing an unknown raised to a different power, cannot usually be solved without an electronic calculator. However, if the powers are 1 and 2, such equations, called *quadratic equations*, can be solved without a calculator. All quadratic equations can be put in the form

$$ax^2 + bx + c = 0,$$

where a, b, and c are given numbers and x is an unknown to be solved for. An example of a quadratic equation might be

$$3x^2 - 4x + 7 = 0.$$

In this example, $a = 3$, $b = -4$, and $c = 7$.

Let's find the solution to the general quadratic equation. First divide by a and then add and subtract $(b/2a)^2$, getting

$$x^2 + \frac{bx}{a} + \left(\frac{b}{2a}\right)^2 - \left(\frac{b}{2a}\right)^2 + \frac{c}{a} = 0.$$

We can rewrite this as

$$\left(x + \frac{b}{2a}\right)^2 - \left(\frac{b}{2a}\right)^2 - \frac{c}{a}.$$

Taking the square root of this equation gives

$$x + b/2a = +[(b/2a)^2 - c/a]^{1/2},$$

and

$$x + b/2a = -[(b/2a)^2 - c/a]^{1/2}.$$

Note that when a square root is taken, there are *two* solutions, differing by a minus sign. These two solutions, one positive and one negative, can be abbreviated by the sign $\pm$; for example, 3 ± 0.5 means $3 + 0.5$ and $3 - 0.5$. Finally, subtracting $b/2a$ from both sides, we get

$$x = -b/2a \pm [(b/2a)^2 - c/a]^{1/2}.$$

This result is often seen in the equivalent form:

$$x = \frac{-b \pm \sqrt{b^2 - 4ac}}{2a}. \tag{A-1}$$

As an example of solving a quadratic equation, consider the equation $2x^2 + 5x + 3 = 0$. Here $a = 2$, $b = 5$, and $c = 3$. Substituting these values into the general solution for x, we get

$$x = -5/4 \pm [25/16 - 3/2]^{1/2} = -5/4 \pm 1/4,$$

or $x = -1$ and $x = -3/2$. It is easy to verify that these are the correct solutions to the original equation.

5. GEOMETRY

a. Angles of a Triangle

The sum of the angles of a triangle add up to 180 degrees. Thus, in Fig. A–1,

$$a + b + c = 180°. \tag{A-2}$$

b. Similar Triangles

If two triangles have the same angles, they are called similar triangles. One triangle is just a blown up version of the other. The two triangles in Fig. A–2, for example, are similar triangles; they have the same angles a, b, and c. The triangle on the right is 1.5 times the size of the one on the left, meaning that each of its sides is 1.5 times the length of the corresponding side of the triangle on the left.

Since each side is multiplied by the same factor in going from one triangle to a similar triangle, the ratio of the lengths of two sides of a triangle is equal to the ratio of the lengths of the corresponding two sides of a similar triangle. In the triangle on the left in Fig. A–2, for example, the ratio of the length of the side opposite the angle a to the side opposite the angle c is f/e. For the similar triangle on the right, the ratio of the length of the side opposite the angle a to the side opposite the angle c is $1.5f/1.5e = f/e$. This is an important result for similar triangles.

c. Ratio of Sides of Right Triangles

A right triangle is a triangle one of whose angles is a right angle, that is, 90 degrees. If either of the other two angles of a right triangle is specified, then the third angle is also determined, since the sum must add to 180 degrees. Therefore, if two right triangles have another angle that is the same (in addition to the right angle), then all three angles are the same, and the two triangles are similar, as shown in Fig. A–3.

It follows that once a single additional angle (other than the right angle) of a right triangle is specified, the ratio of lengths of the sides of that triangle can depend only on that angle. The only other thing the ratio of sides might depend on is the overall size of the triangle, but any other right triangle with

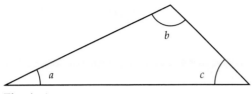

Fig. A–1

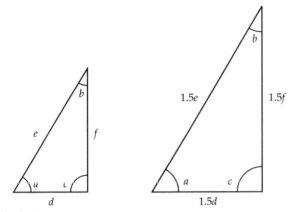

Fig. A–2

the same additional angle would be similar to the first triangle, and we have already established that the ratio of any two sides of similar triangles is the same, independent of their overall size.

In the right triangle in Fig. A–4, the ratios y/x, y/z, and x/z are completely determined once the angle a is given. For example, if $a = 30°$, then $y/x = 0.5774$, $y/z = 0.5$, and $x/z = 0.866$. If $a = 48°$, $y/x = 1.1106$, $y/z = 0.7431$, and $x/z = 0.6691$. For any value of a, these ratios can be gotten from an electronic calculator.

The ratios of the lengths of the sides of a right triangle have names. Consider the right triangle in Fig. A–4. The side opposite the 90 degree angle, labeled z, is called the *hypotenuse*. The ratio of the length of the side opposite the angle a to the length of the hypotenuse, y/z, is called the *sine* of a and is abbreviated sin a; the ratio of the length of the side adjacent to the angle a to the hypotenuse, x/z, is called the *cosine* of a and is abbreviated cos a. The ratio y/x is called the *tangent* of a and abbreviated tan a.

The sine, cosine, and tangent of any angle may be looked up in a book or computed on an electronic calculator. For example, tan 15° = 0.2679, sin 15° = 0.2588, and cos 15° = 0.9659; tan 30° = 0.5774, sin 30° = 0.5, and cos 30° = 0.866; tan 45° = 1.0, sin 45° = 0.7071, and cos 45° = 0.7071; tan 60° = 1.7321, sin 60° = 0.8660, and cos 60° = 0.5; tan 75° = 3.7321, sin 75° = 0.9659, and cos 75° = 0.2588; tan 90° = *infinity*, sin 90° = 1.0, and cos 90° = 0.0.

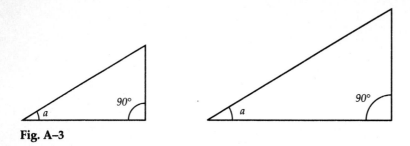

Fig. A–3

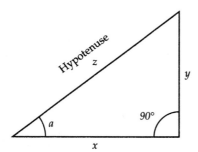

Fig. A–4

d. The Pythagorean Theorem

The lengths of the sides of a right triangle have an important relationship, called the *Pythagorean theorem*. Consider again the right triangle shown in Fig. A–4. Then, the Pythagorean theorem states

$$x^2 + y^2 = z^2. \tag{A–3}$$

This relationship between the sides of a right triangle is one of the most fundamental results in geometry. Although the Pythagorean theorem carries the name of the Greek mathematician Pythagoras (ca. 560–ca. 480 B.C.), it was known by Babylonian mathematicians a thousand years earlier.

APPENDIX B

The Second Law of Thermodynamics and the Behavior of Systems with Large Numbers of Molecules

In the main text we worked out the behavior of systems having a very small number of molecules. Although the trends we discovered—the winding down of a pendulum and the flow of heat from a hot body to a cold one—were correct, they were not strong trends because the number of molecules we used in our examples was unrealistically small. In this Appendix, we extrapolate to large numbers of molecules. The mathematics is not difficult, but you will have to think carefully. Here we go.

1. NUMBER OF STATES FOR A GAS OF MANY MOLECULES

It is difficult to derive the exact number of states for a realistic gas having many molecules, but we can derive good approximations. Assume first that the number of units of energy in the gas E_g is much larger than the number of molecules N so that, on average, every molecule has some energy. For the moment, pretend that the molecules are *distinguishable* and can be labeled; we will make them indistinguishable later. Then, a state of the gas is specified by giving the energy of each molecule $(E_1, E_2, E_3, ..., E_N)$, where E_1 is the energy of the first molecule, E_2 is the energy of the second molecule, and so on, and the dots refer to all the molecules between the third and the last. The individual molecular energies must sum up, of course, to the total energy in the gas:

$$E_1 + E_2 + E_3 + \cdots + E_N = E_g. \tag{B-1}$$

How many different ways are there for the energy E_g to be distributed among the N molecules? Let's consider a typical molecule. It can have an energy of 1 unit or 2 or 3, on up to E_g. (As in Chapter II, we will assume that each molecule can have only a whole number of units of energy. This assumption simplifies the discussion and does not limit the validity of our results.) The average energy of a molecule is E_g/N. If our typical molecule has an energy much larger than the average, then many of the remaining molecules won't have any energy at all, and the result will be a state with a lot of zeros, like $(E_1,0,0,E_4,E_5,0,0,...,E_N)$.

There aren't nearly as many such states as those in which most of the molecules have at least a little energy. For example, in a three-molecule gas with 4 units of energy there are only 3 states in which two molecules have zero energy: (4,0,0), (0,4,0), and (0,0,4), as compared to 12 states in which one or none of the molecules has zero energy: (3,1,0), (3,0,1), (1,0,3), (1,3,0), (0,3,1), (0,1,3), (2,2,0), (2,0,2), (0,0,2), (1,1,2), (2,1,1), and (1,2,1). (Remember that the molecules are distinguishable for the time being.)

Thus, for each value of E_g, the greatest number of different states of the system is produced when our typical molecule has an energy at or below the average E_g/N, that is, 1 unit or 2 or 3, on up to E_g/N, for a total of E_g/N different values of energy. If the typical molecule can have E_g/N different values of energy, and there are N molecules, then the total number of different combinations for the system is $(E_g/N)^N$. This is analogous to the two dice, where each die can have 6 values and the total number of different combinations for the two dice is $6 \times 6 = 6^2$.

Finally, we have to remember that the molecules are really indistinguishable. They cannot be labeled. Therefore, the approximate formula we have just derived *overcounts* the number of states. For every state, we have to go back and eliminate all the states that could be obtained from that state by rearranging the molecules. Given the N energies $E_1, E_2, E_3, \ldots, E_N$, how many different arrangements of molecules are there? The first molecule can take any one of the N energies. For each of these values of energy, the second molecule can take any one of the remaining $N - 1$ energies. For each of these, the third molecule can take any of the remaining $N - 2$ energies. And so on. Thus the total number of different arrangements is $N(N - 1)(N - 2) \ldots 1$. A shorthand notation for this is $N!$, that is,

$$N! = N(N - 1)(N - 2) \ldots 1. \tag{B-2}$$

For example, $4! = 4 \times 3 \times 2 \times 1 = 24$. Thus we have counted $N!$ different states for every state that we should have counted only once. We have to divide our previous result by $N!$. Our final estimate, then, for the number of states corresponding to a total energy E_g divided among N indistinguishable molecules is

$$\Omega = \frac{(E_g / N)^N}{N!}.$$ (B–3)

As in Chapter II, we use the symbol Ω to denote the number of states. As in the case of the dice or the three-molecule gas, the probability of a configuration having any particular amount of energy E_g is proportional to the number of states Ω for that value of E_g. The important feature of Eq. (B–3) is the way that Ω depends on E_g, varying as E_g^N. This relationship between Ω and E_g turns out to be approximately true for a wide variety of systems, much more complicated than the ones we have considered.

2. BEHAVIOR OF A PENDULUM IN A REALISTIC GAS

Realistic gases have many, many molecules. We now want to use Eq. (B–3) to estimate the probability of finding a pendulum in a realistic gas with some sizable fraction of the total energy, say 3/11 or more. Remember that this probability was about 50% for the three-molecule gas. We obtained it by summing the probabilities of all configurations with 3/11 or more of the total energy in the pendulum. For systems with an enormous number of molecules, there would be an enormous number of such configurations, and the sum would be difficult to compute.

However, there is a shortcut. Note from Table II–3 that the probability that a pendulum in a three-molecule gas has 3/11 or more of the total energy is close to the *ratio* of the probability of the configuration with 3/11 of the energy in the pendulum to the probability of the configuration with none of the energy in the pendulum (the most probable configuration), $(10/83)/(16/83) = 10/16$. For large numbers of molecules, this ratio becomes an excellent estimate of what we want.

Now, apply Eq. (B–3), our approximate formula for large N. Remember that when the pendulum has 3/11 of the total energy, the gas has 8/11 of the total energy, and when the pendulum has none of the total energy, the gas has all of the total energy. Thus, we want to take the ratio of Ω for $E_g = 8/11\, E_{tot}$ to Ω for $E_g = E_{tot}$:

$$\frac{\Omega_{E_g = 8/11\, E_{tot}}}{\Omega_{E_g = E_{tot}}} = (8/11)^N.$$ (B–4)

Note the cancellations. This ratio is a good approximation to the probability that the pendulum will have 3/11 or more of the total energy. It decreases rapidly as the number of molecules N increases. For a three-molecule gas, it gives $(8/11)^3 = 0.38$, in fair agreement with our exact calculation. What does it give for a realistic gas, with a large value of N? A cubic foot of air has about 10^{24} molecules. Substituting $N = 10^{24}$ into the above formula gives an unimaginably small probability for the pendulum to have 3/11 or more of the total energy. Thus, there is virtually no chance that a pendulum at rest in a realistic box of air could start swinging with a substantial fraction of the total energy.

We can estimate what fraction of the total energy we *would* expect the pendulum to pick up, starting from rest. For the three-molecule gas, we found that the probability was about 50% that a pendulum initially at rest, or in any other initial state, would later have 3/11 or more of the total energy, and we interpreted this to mean that the pendulum would eventually spend about half the time with this much energy or more. Equivalently, the pendulum in the three-molecule gas spends about half its time with less than 3/11 of the total energy. This energy, therefore, serves as a good indicator of the expected energy of the pendulum at any moment.

For the realistic case with many gas molecules, we want to calculate the energy E_* such that the pendulum would have less than this energy half the time. In other words, we want to find the value of E_* such that the ratio of the probability of the configuration with $E_g = E_{tot} - E_*$ to the probability of the configuration with $E_g = E_{tot}$ is equal to 1/2. Using Eq. (B–3), we have the condition

$$\frac{1}{2} = \frac{\Omega_{E_g = E_{tot} - E_*}}{\Omega_{E_g = E_{tot}}} = \frac{(E_{tot} - E_*)^N}{E_{tot}^N} = \left(1 - \frac{E_*}{E_{tot}}\right)^N, \qquad \text{(B–5)}$$

which is an equation to be solved for E_*/E_{tot}. From simple algebra, the solution is

$$\frac{E_*}{E_{tot}} = 1 - \left(\frac{1}{2}\right)^{1/N}. \qquad \text{(B–6)}$$

Recall that E_* is the "expected" energy of the pendulum: the pendulum will have an energy less than E_* half the time. Notice that as N increases, E_*/E_{tot} gets smaller and smaller; that is, with an increasing number of molecules, the fraction of the total energy expected in the pendulum gets smaller and smaller. For $N = 3$, Eq. (B–6) gives $E_*/E_{tot} = 0.21$, in fair agreement with our exact result of 3/11. For $N = 5$, 100, and 10,000, Eq. (B–6) gives $E_*/E_{tot} = 0.13$, 0.007, and 0.00007, respectively, showing how E_*/E_{tot} decreases with increasing N.

A good approximation to our formula for large N is

$$\frac{E_*}{E_{tot}} = \frac{0.7}{N},$$

which says that the system will evolve to a condition in which the pendulum has about the same amount of energy as an average molecule, E_{tot}/N, just as if the pendulum were one of the molecules. *The evolution tends toward a condition in which the total energy of the system is equally distributed among all of its parts.* This is, in fact, an alternative statement of the second law of thermodynamics.

For a cubic foot of air (0.02 cubic meters), with $N = 10^{24}$, the above result gives $E_*/E_{tot} = 0.7 \times 10^{-24}$, which is a very tiny fraction of the total energy indeed. This fraction of the heat energy in 0.02 cubic meters of gas at room temperature would nudge a 1-pound pendulum upward about 10^{-19} centimeters, much less than the width of an atom. Thus, for practical purposes, a pendulum started at rest in a realistic gas will remain at rest; a pendulum started with a large fraction of the total energy will wind down until it is at rest.

3. HEAT FLOW AND EQUALIZATION OF TEMPERATURES IN REALISTIC SYSTEMS

Our simple system of two 3-molecule gases shows the correct *trend* of evolution toward equalizing the temperatures. But, just as in the case of the pendulum and the three-molecule gas, large departures from the trend are not unlikely. Such departures become extremely rare for systems with large numbers of molecules.

To get a feeling for more realistic systems, we can again use our approximate formula for a large number of molecules, Eq. (B–3). Let there be N_A molecules in gas A and N_B molecules in gas B. Then, from Eqs. (II–3) and (B–3), an approximation to the total number of states available to the system is

$$\Omega_{tot} = \Omega_A \times \Omega_B = \frac{(E_A / N_A)^{N_A}}{N_A!} \times \frac{(E_B / N_B)^{N_B}}{N_B!}. \qquad (B–7a)$$

Now, what we want to do is to find the values of E_A and E_B for which Ω_{tot} is *largest*. These values will represent the most probable configuration of the system, since probability is just proportional to the number of states.

First, separate the quantities that do not change for a given system. The numbers of molecules N_A and N_B are fixed for a given system, as is the total energy E_{tot}. Then, substituting $E_{tot} - E_A$ for E_B and dividing and multiplying by $E_{tot}^{N_A} E_{tot}^{N_B}$, Eq. (B–7a) can be rewritten as

$$\Omega_{tot} = \left[\frac{(1/N_A)^{N_A}}{N_A!} \frac{(1/N_B)^{N_B}}{N_B!} E_{tot}^{N_A + N_B} \right] \left(\frac{E_A}{E_{tot}} \right)^{N_A} \left(1 - \frac{E_A}{E_{tot}} \right). \qquad (B–7b)$$

For a given system, everything inside the brackets is fixed; the only quantity that varies for different divisions of the total energy is the ratio E_A/E_{tot}. (Of course, the ratio E_B/E_{tot} also varies, but this variation is completely determined by E_A/E_{tot}, since $E_B/E_{tot} = 1 - E_A/E_{tot}$.)

Using Eq. (B–7b), we can first ask what is the most probable configuration. In other words, for what value of the ratio E_A/E_{tot} is Ω_{tot} the largest? The ratio E_A/E_{tot} can vary between 0 and 1. At both of these extremes, $\Omega_{tot} = 0$, as can be seen from Eq. (B–7b). For intermediate values of E_A/E_{tot}, Ω_{tot} is bigger than 0, and for one of these intermediate values, Ω_{tot} is largest. To simplify the math, let $x = E_A/E_{tot}$ and $y = N_A/N_B$, and raise Eq. (B–7b) to the $1/N_B$ power. We then get the equation

$$\Omega_{tot}^{1/N_B} = (\text{constant}) \times x^y(1 - x). \tag{B–7c}$$

The constant involves only N_A, N_B, and E_{tot}, which don't change for a given system. We also know that the value of x that maximizes Ω_{tot}^{1/N_B} also maximizes Ω_{tot}. Thus, we have reduced the problem to finding the value of x that makes the combination $x^y(1 - x)$ the largest.

It can be shown using calculus that $x^y(1 - x)$ is largest for

$$x = \frac{y}{1 + y}. \tag{B–8a}$$

For each given y, this is the value of $x = E_A/E_{tot}$ that maximizes the number of states, and hence probability, of the system. To make the result in Eq. (B–8a) plausible, even if you cannot derive it, you can pick a value of y (which is fixed for a given system) and compute $x^y(1 - x)$ for a lot of values of x, starting from 0 and going up to 1. For example, if you pick $y = 3$ and take x in increments of 0.1, you will get the results of Table B–1. As you can see, $x^y(1 - x)$ is largest for x somewhere between 0.7 and 0.8. This result agrees well with the exact result you get by substituting $y = 3$ into Eq. (B–8a), $x = 0.75$.

TABLE B–1: Relative Number of States in a System of Two Gases with Many Molecules for Various Divisions of the Energy

$y = N_A/N_B = 3$	
$x = E_A/E_{tot}$	$x^y(1 - x)$
0	0
0.1	0.0009
0.2	0.0064
0.3	0.019
0.4	0.038
0.5	0.063
0.6	0.086
0.7	0.10
0.8	0.10
0.9	0.073
1	0

We now put our result, Eq. (B–8a), into a more obvious form. First, multiply both sides of Eq. (B–8a) by $(1 + y)$ and solve for y: $y = x/(1 - x)$. Now, multiply this last equation by $(1 - x)/y$ to get

$$\frac{x}{y} = 1 - x.$$

If we now go back to the quantities that x and y stand for, we get the result

$$\frac{E_A / E_{tot}}{N_A / N_B} = 1 - \frac{E_A}{E_{tot}},$$

or, dividing by N_B,

$$\frac{E_A / E_{tot}}{N_A} = \frac{(1 - E_A / E_{tot})}{N_B}. \tag{B–8b}$$

Multiply both sides of Eq. (B–8b) by E_{tot}, substitute $E_B = E_{tot} - E_A$, and use the definition of temperature, $T_A = E_A/N_A$, $T_B = E_B/N_B$, to obtain the equivalent condition

$$T_A = T_B. \tag{B–8c}$$

This is our final result. The configuration of maximum probability is the one for which the two gases have equal temperatures, in agreement with our exact treatment of two 3-molecule gases. As the system naturally evolves toward configurations of greater probability, it will gradually equalize the temperatures of its two parts, with the average energy of a molecule in gas A equal to that of a molecule in gas B.

Finally, we can ask how far the system is likely to wander from the condition of equal temperatures. Even in a system of many molecules, the system will occasionally depart from its most probable configuration because probabilities are, after all, only statements about average behaviors.

Suppose that the two gases have been in contact for a very long time, so that there has been plenty of time for them to have come to the same temperature. We now calculate the range of E_A/E_{tot} such that the system can be found in this range half the time. This calculation is analogous to our calculation of the energy E_* for the pendulum. For simplicity, let us take $N_A = N_B = N$. Then, Eq. (B–8b) shows that Ω_{tot} is largest for $E_A/E_{tot} = 1/2$. This is the configuration of maximum probability. Let E_*/E_{tot} be the value of E_A/E_{tot} such that the system can be found half the time with E_A/E_{tot} between E_*/E_{tot} and $1/2$, the most probable value. When gas A has energy E_*, gas B has energy $E_{tot} - E_*$. These values can be substituted into Eq. (B–7a) with $N_A = N$ and $N_B = N$. By analogy with the pendulum problem, we find E_* by solving

$$\frac{1}{2} = \frac{\Omega_{E_A = E_*}}{\Omega_{E_A = E_{tot}/2}} = 2^{2N} \left(\frac{E_*}{E_{tot}} \right)^N \left(1 - \frac{E_*}{E_{tot}} \right)^N. \tag{B–9a}$$

Raising this equation to the $1/N$ power and dividing by 4, we get

$$\left(\frac{E_*}{E_{tot}} \right) \left(1 - \frac{E_*}{E_{tot}} \right) = \frac{2^{-1/N}}{4}, \tag{B–9b}$$

or

$$-\left(\frac{E_*}{E_{tot}} \right)^2 + \frac{E_*}{E_{tot}} - \frac{2^{-1/N}}{4} = 0. \tag{B–9c}$$

The roots of this quadratic equation for the unknown E_*/E_{tot} are

$$\frac{E_*}{E_{tot}} = \frac{1 \pm (1 - 2^{-1/N})^{1/2}}{2}. \tag{B–9d}$$

Here, we can easily understand that the two roots correspond to values above and below the most probable value of $1/2$. We can also see that as N gets larger, E_*/E_{tot} gets closer and closer to $1/2$; that is, the system wanders less and less from its most probable configuration. For example, for $N = 5$, 100, and 10,000, Eq. (B–9d) gives $E_*/E_{tot} - 1/2 = \pm 0.18$, 0.042, and 0.0042, respectively.
 A good approximation for large N is

$$\frac{E_*}{E_{tot}} - \frac{1}{2} = \pm \frac{0.42}{N^{1/2}}. \tag{B–10}$$

For gases of familiar sizes and densities, say with $N = 10^{24}$, the random departure from the most probable configuration, once this configuration has been achieved, is extremely small. For all practical purposes, the system will smoothly evolve toward the condition in which the temperatures of its parts are equal and will remain there.

Index